OFFICIALLY
DISCARDED

Interstellar dust grain: diameter 4×10^{-5} inch

Bacterium: diameter 4×10^{-5} inch

Black hole: diameter 40 miles

Large moon crater: diameter 120 miles

Largest asteroid: diameter 620 miles

Mars: diameter 4,217 miles

White dwarf: diameter 5,000 miles

Venus: diameter 7,521 miles

THE SUN

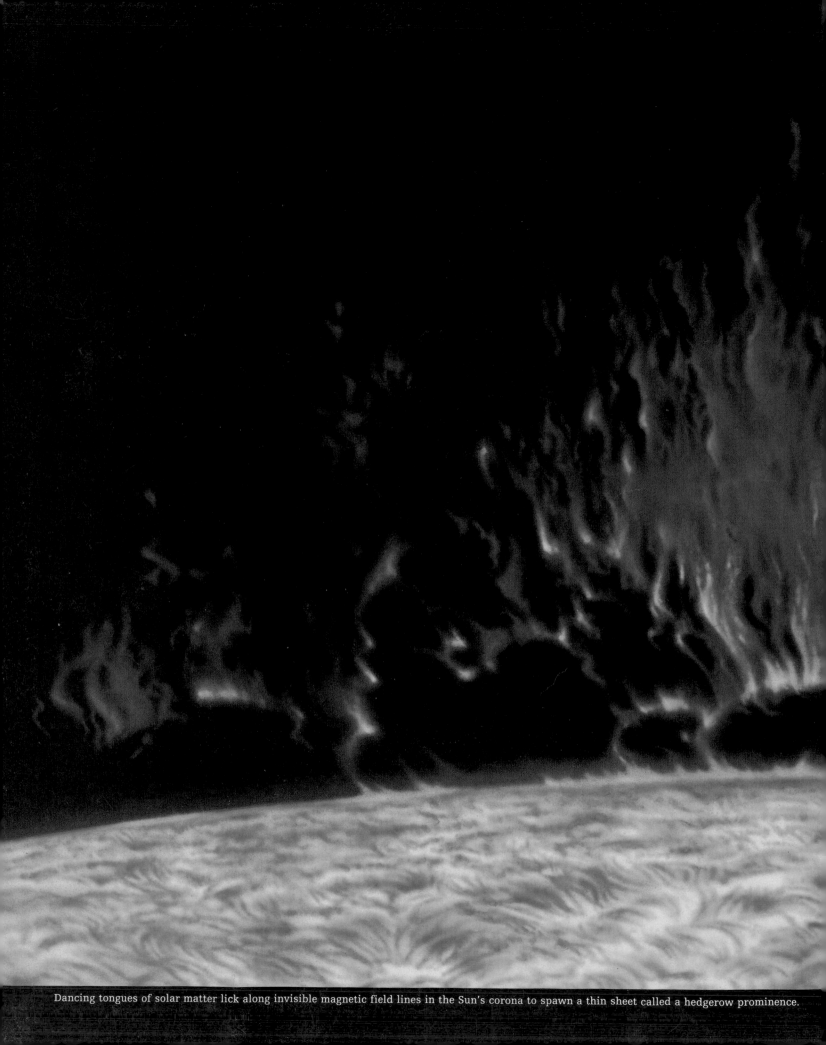

Dancing tongues of solar matter lick along invisible magnetic field lines in the Sun's corona to spawn a thin sheet called a hedgerow prominence.

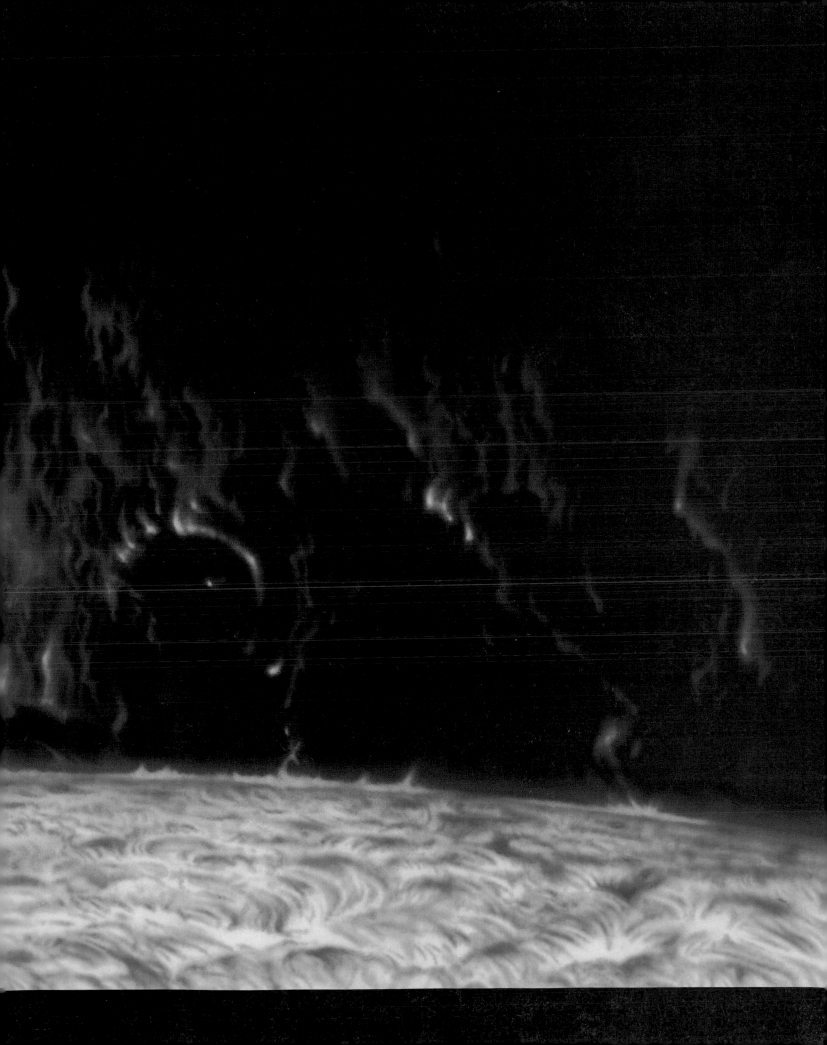

®

Other Publications:
AMERICAN COUNTRY
THE THIRD REICH
THE TIME-LIFE GARDENER'S GUIDE
MYSTERIES OF THE UNKNOWN
TIME FRAME
FIX IT YOURSELF
FITNESS, HEALTH & NUTRITION
SUCCESSFUL PARENTING
HEALTHY HOME COOKING
UNDERSTANDING COMPUTERS
LIBRARY OF NATIONS
THE ENCHANTED WORLD
THE KODAK LIBRARY OF CREATIVE PHOTOGRAPHY
GREAT MEALS IN MINUTES
THE CIVIL WAR
PLANET EARTH
COLLECTOR'S LIBRARY OF THE CIVIL WAR
THE EPIC OF FLIGHT
THE GOOD COOK
WORLD WAR II
HOME REPAIR AND IMPROVEMENT
THE OLD WEST

This volume is one of a series that
examines the universe in all its aspects,
from its beginnings in the Big Bang to the
promise of space exploration.

VOYAGE THROUGH THE UNIVERSE

THE SUN

BY THE EDITORS OF TIME-LIFE BOOKS
ALEXANDRIA, VIRGINIA

CONTENTS

1/The Face of the Sun

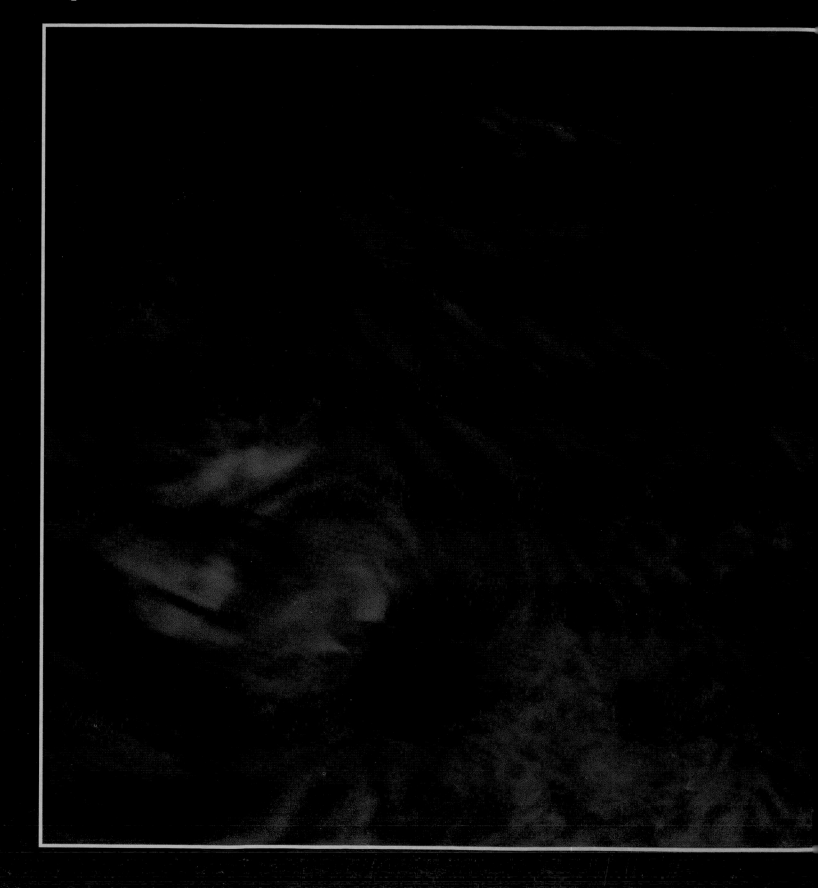

called prominences and punctuated by dark sunspots, the turbulent Sun emerges in an image taken by an instrument tuned to hydrogen alpha, a wavelength of light characterized by a vibrant red hue.

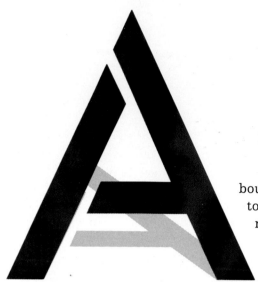bout 30,000 years ago, a process crucial to all life on the planet today was carried out tens of trillions of times per second deep in the core of the Sun. Crushed together under pressure of the star's enormous mass, four hydrogen atoms underwent a series of complex fusion reactions whose end products included a single atom of helium and several packets of energy, primarily in the form of very high frequency, short-wavelength gamma rays. As the gamma ray packets began to migrate toward the solar surface, their progress was slowed by innumerable collisions with the nuclei and free electrons of the fantastically hot, dense solar gases. Each collision reduced the frequency of the gamma rays' vibration, gradually shifting them to x-ray and ultraviolet wavelengths, and then to the wavelengths of visible light. About eight minutes ago, the packets finally broke the surface of the Sun and sped outward into space. The sunlight now striking Earth delivers energy that was created when humans were fending off saber-toothed tigers with stone weapons.

Less than half a billionth of the Sun's energy output bathes the planet each second, yet even those few crumbs from the solar table are enough to nourish and power the entire globe. As heat absorbed at the equator is transported toward the cooler poles by circulating air currents, ocean water evaporates to form clouds and rain; vast rivers in the sea—the Gulf Stream in the Atlantic Ocean, for example, and the Peru Current in the Pacific—moderate the climate of the shores they flow past. The shower of sunlight also triggers photosynthesis in the chlorophyll molecules of green leaves, a reaction that produces carbohydrates, the basis for all terrestrial life *(pages 91-99)*. Carbohydrates get passed up the food chain, nurturing herbivores, then carnivores and such omnivores as human beings. Plants that escape this cycle of consumption ultimately yield their stored solar energy as well: The fossil fuels now running oil heaters, factories, and cars are nothing more than the remains of ancient foliage, transmuted through millennia of compression below Earth's surface.

Reliable as the Sun appears, it may yet prove an inconstant source of energy. Scientists interpreting data collected since 1980 by NASA's Solar Maximum Mission satellite have concluded that the Sun dims slightly—by about 0.1 percent—whenever a large group of the cool, dark surface blemishes known as sunspots march across its countenance. On average, however (and counter to expectations), the Sun brightens as the eleven-year cycle of sunspot activity increases and grows dimmer as the nadir of that cycle, called solar minimum, approaches.

A panoply of additional traits testify to the power and unpredictability of Earth's nearest star. Solar flares belch fiery gases by the billions of tons, and incandescent prominences soar and loop as much as two million miles above the surface, sometimes hurling gas into interplanetary space, sometimes reeling it back by means of magnetism. From openings in the magnetic field of its outermost atmospheric layer—a hot, tenuous region called the corona—the Sun disgorges a mighty stream of highly charged particles, a "solar wind" that buffets the entire Solar System. At Earth, for example, the wind's energized particles continually spiral into the planet's atmosphere to illuminate the night skies with colorful auroral displays. And when a flare causes the solar wind to blow especially strong, the result can be power-grid failures that plunge whole sections of a country into darkness.

PICTURES OF A STAR

Long before scientists truly understood the ways in which Earth depends on the Sun, various ancient cultures—including the Egyptians, the Greeks, and the Romans—revered the blazing orb as a god, a benign provider of light and life. Amid the oak groves of Great Britain and Gaul, Celtic priests performed rituals to revive the autumnal Sun, cutting green plants such as mistletoe to bring it back to life and lighting fires to restore its strength.

Alongside this instinctive urge to worship, there grew a desire to study and comprehend. Followers of the Greek mathematician Pythagoras, for example, concluded in the sixth century BC that Earth revolved about a "central fire" every twenty-four hours; according to the Pythagoreans, Earth was shielded from this central fire by a nearby orbital partner called the counter-Earth but periodically exposed to the distant—and stationary—Sun. Within two centuries, however, Aristotle had converted most early astronomers to his view that Earth was the unmoving center of everything and that all lesser bodies,

Matched against the vastness of the Sun, Earth dwindles to a small dot in the comparative rendering below. Although the Sun's dimensions are merely average by stellar standards, its diameter—864,945 miles—is 110 times that of humankind's home planet.

the Sun included, danced attendance around it. (Aristotle also taught that the Sun was made of pure fire, free of imperfections.) This geocentric outlook would prevail until the Renaissance, when the Polish cleric Nicolaus Copernicus established that Earth revolves around the Sun.

While one school of astronomers tried to fix the Sun's place in the heavens, another attempted to determine its composition. In the fifth century BC, Anaxagoras—a Greek philosopher and astronomer who had correctly explained solar eclipses—posited that the Sun was "a mass of red-hot metal." The accuracy of that concept had improved little by the eve of the nineteenth century, when British astronomer William Herschel proposed that the Sun was a dark, solid body swathed in luminous clouds and populated by beings "adapted to the peculiar circumstances of that vast globe."

With technological breakthroughs in the nineteenth and twentieth centuries, a more accurate picture of the Sun began to emerge. Solar physicists now know, for instance, that the Sun is a rather commonplace celestial object. If all the stars in the sky were arrayed in a field and proportionately reduced so that the smallest was the size of a grain of sand, the largest would be as

SUN-WATCHING PRECAUTIONS

Solar astronomers spend many hours gazing at the Sun, but they always take care to protect their eyes. For example, they never look at the Sun directly; even a momentary glance can be painful and could do permanent damage to vision. Telescopes and binoculars pose an even greater threat. These instruments concentrate the Sun's rays much as a magnifying glass does; exposure to the intensified heat and brightness would almost certainly cause blindness.

The best way for nonprofessionals to study the Sun in detail is with the technique called projection (opposite). The observer should shade the larger-diameter lens of a telescope or one lens of a pair of binoculars with a cardboard collar. (The second binocular lens should be taped shut with a piece of cardboard.) The solar image can then be projected by pointing the lens at the Sun—again, taking care never to sight through the instrument—and focusing the beam onto a sheet of white paper in a dark room. A surprising number of details will emerge from an image so projected, among

them sunspots and the bright patches known as plages.

For direct scrutiny of the Sun, the only device that successfully screens out dangerous solar rays is a dark filter, available for amateur telescopes. Hydrogen alpha filters, which block light from the photosphere at every wavelength except that of hydrogen alpha, permit viewing of the chromosphere and the plages, filaments, prominences, and flares that arise there. (Amateurs may wish to correlate their results with those of professional astronomers from the American Association of Variable Star Observers in Cambridge, Massachusetts.)

Eclipses, too, require precautions. Although the Moon blocks out a portion of the Sun or obscures the entire solar orb, enough of the hot surface remains exposed to make it extremely unsafe to look at an eclipse even through x-ray film or dark glass. Instead, a pinhole projector will provide safe viewing: Cover one end of a cardboard tube with aluminum foil, then pierce the foil with a needle and project the eclipse through this tiny aperture onto a sheet of white paper.

big as a geodesic dome 300 feet high. In this stellar gathering, the Sun, along with most other stars, would be the size of a basketball.

Of ordinary dimensions, the Sun is also of ordinary brightness; if it were not so near, other stars would easily outshine it. The North Star, Polaris, shines about 6,000 times brighter than the Sun, and the star Rigel in the constellation Orion emits 15,000 times more energy. In mass, too, Earth's home star is average; most stars possess a mass that is between 0.1 and 10 times that of the Sun.

Unspectacular it may be in universal terms, but to observers on Earth the Sun remains an object of stupendous proportions. With a diameter of nearly 865,000 miles, this fiery ball of superheated hydrogen and helium gases contains 99.9 percent of all matter in the Solar System. A million Earths could fit inside the Sun, with room to spare.

Scientists are just beginning to appreciate the complexity of its behavior. They have discovered, for example, that the Sun's core may be rotating at a different rate from its surface and that the entire star may be vibrating like

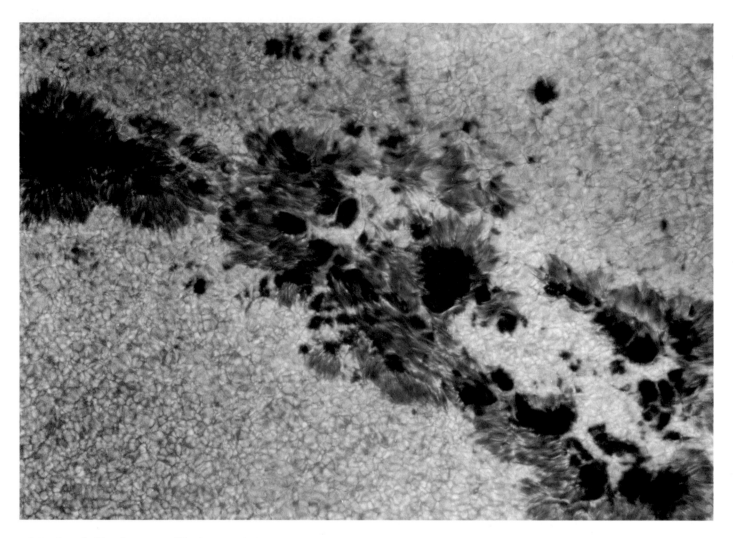

a big, hot bell. These oscillations originate at all levels of the Sun's complex architecture, from the convection zone just below the surface, to the radiative zone that occupies the middle layers, to the unimaginably fiery core generating temperatures of 15.5 million degrees Kelvin.

As well as gathering the Sun's vital statistics and gauging its behavior, solar physicists believe they have successfully traced its origins. The star began to form, they theorize, some 4.6 billion years ago, when shock waves from a supernova explosion caused the center of a nearby cosmic dust cloud to collapse into a dense core. As additional gas and dust spiraled into this core, gravity drove the internal temperature and pressure steadily higher until, at about 11 million degrees Kelvin, thermonuclear fusion got under way. Fusion reactions have powered the Sun ever since, and should continue to do so for at least another 5 billion years.

The only portion of this middle-aged star visible from Earth is its atmosphere, a solar envelope that consists of three layers of distinctly different

Surrounded by bright granules of superheated gas rising from the solar interior, a cluster of sunspots trails across the photosphere, the Sun's visible layer. Because sunspots are easily detected by earthbound observers, astronomers have recorded their comings and goings for centuries. In this chain, which stretched 55,000 miles across the solar surface on November 15, 1970, each of the larger spots is about the size of Earth.

temperatures and behavior. Possessing only optical means of observing the Sun, early astronomers therefore examined these three layers first. The innermost was the easiest to distinguish. Often referred to as the Sun's surface, it is a highly luminous, 200-mile-deep region called the photosphere, or sphere of light, whose hydrogen and helium gases glow at temperatures of 6,000 to 10,000 degrees Kelvin. So intensely bright is the photosphere that it ordinarily obscures the two additional layers—the chromosphere and the corona—that rise above it.

Not until the development of the spectroscope, which separates the constituent wavelengths of sunlight for detailed analysis, and the coronagraph, which masks the photosphere so that the chromosphere and corona can emerge in silhouette around it, did astronomers succeed in viewing the upper atmospheric tiers in the absence of a total solar eclipse. Such tools afforded the first glimpses of the quirky and violent nature of the Sun's outer atmosphere. The chromosphere—so named because it glows fiery red when seen through a spectroscope—is almost twice as hot as the photosphere and serves as the staging area for some of the Sun's most spectacular displays. The corona has temperatures higher than any part of the Sun except its center and forms a shimmering, tenuous veil that extends at least two million miles outward from the roof of the chromosphere.

SOLAR BLOTCHES

Because early astronomers focused most of their attention on the photosphere, sunspots became the first extensively studied solar phenomenon. Theophrastus of Athens reported seeing blotches on the Sun in 300 BC, and the Chinese astronomer Kan Te began recording his observations of the spots at about the same time. Although the dust storms that sweep across the north China plains obscured the stars by night, they dimmed the Sun just enough by day that observers could view it—and its spots—without damaging their eyesight. Between 28 BC and AD 1638, Chinese astronomers witnessed 112 outbreaks of sunspots.

Unacquainted with this record but familiar with the doctrines of Aristotle, Italian astronomer Galileo Galilei may have expected to find a uniform sphere of brightness when he began observing the Sun with the newly invented telescope in 1610. What he observed instead was a series of small, dark spots that made a stately procession across the disk, completing their transit in about two weeks' time. (Much later, astronomers would interpret this as evidence that the Sun's surface is rotating.) Although Galileo's findings were branded as heresy by a Church that had adopted the Aristotelian precept of solar perfection, dozens of observers recorded sunspots in the decades after Galileo's discovery.

The first astronomer to discern a larger rhythm underlying the behavior of sunspots was an amateur. As a pharmacology student at the University of Berlin from 1810 to 1812, Heinrich Schwabe had developed an interest in the natural sciences, especially botany, mineralogy, and astronomy. He might have remained forever a dilettante observer but for a stroke of fortune: In 1825, having returned to his hometown of Dessau to run the family's apothecary, Schwabe won a high-quality telescope in a local lottery, and during lulls in the pill-and-potion business he used the instrument to view the Sun.

Four years later, inspired by accounts of William Herschel's 1781 discovery of Uranus, Schwabe sold his pharmacy and set out to find a planet of his own. Various of his contemporaries had postulated the existence of a planet called Vulcan, which was supposed to have an orbit inside that of Mercury, and Schwabe intended to catch the fugitive body in the act of making regular passages across the Sun's face. In order to accomplish this, however, he would first have to differentiate the dark blotch that would be a planet passing in front of the Sun from sunspots. The quest produced a serendipitous result. In 1843, after seventeen diligent years of observing the Sun on every cloudless day, Schwabe announced that the number of sunspots rose and fell in a steady cycle of 10.4 years (corrected by later astronomers to 11 years). The period of greatest activity has since come to be known as the sunspot maximum, the quietest period the sunspot minimum. Although he had failed to find Vulcan, Schwabe was well pleased. "I may compare myself to Saul," he proclaimed, "who went to seek his father's ass and found a Kingdom."

Observing the Sun from 1853 to 1861, another amateur British astronomer, Richard Carrington, discovered further predictability in sunspots: Not only do they come and go at regular intervals, but they migrate toward the Sun's equator in systematic fashion.

Carrington, who was born into a wealthy family of brewers in 1826, began his studies at Trinity College in the topic of his father's choice, theology. While there, he happened to attend a series of astronomy lectures by the director of the Cambridge Observatory, Professor James Challis, and the subject so captivated him that he switched allegiances. After graduating in 1848, Carrington joined the University of Durham's observatory. Soon he was chafing at the paltry sums allotted to the purchase of better instruments. Astronomers throughout the ages have known this frustration, but Carrington's personal fortune allowed him to do something about it: In 1853, he built a magnificently equipped observatory at a place called Redhill.

During eight years of solar observations from Redhill, Carrington kept meticulous, day-by-day records of sunspot locations. In a paper published in 1863, he showed that the spots originate in two broad bands on the Sun, one occupying the region between 20 and 40 degrees north of the solar equator and the other occupying the same latitudes south of the equator. (Later astronomers revised these figures to between 30 and 35 degrees solar latitude.) Carrington further demonstrated that in the course of an eleven-year sunspot cycle, the sunspots' latitude of origin gradually drifts toward the equator until only a very few—and then none—emerge in the narrow band between 0 and 5 degrees. As the spots fade out in the Sun's lower latitudes, Carrington noticed, the pattern begins to repeat itself: Sunspots again break out in the higher latitudes, heralding the start of a new eleven-year cycle, and every new spot thereafter appears in a location slightly nearer the Sun's equator.

Carrington also revealed that sunspots in the higher latitudes take about thirty days to complete one full circuit around the Sun (calculated by doubling the time they take to cross the solar hemisphere visible from Earth), while those near the equator move once around in only twenty-seven days. This meant that different parts of the Sun's surface rotate about its axis at different speeds, a phenomenon that modern astronomers call differential rotation. In 1861, Carrington fixed the average solar rotation as seen from Earth at 27.28 days; this finding, he hoped, would win him the Cambridge Observatory directorship that Challis was about to vacate. When Carrington lost his bid to John Couch Adams, an astronomer-mathematician who was one of the two independent discoverers of Neptune, he fell into a funk so deep that he sold the Redhill observatory and devoted himself to managing the family brewery. Despite this dramatic break, Carrington could never quite resist the pull of the stars; within four years he had established a new observatory atop a remote hill in Surrey, where he continued to scan the heavens until his death in 1875.

THE SPHERE OF COLOR

Missing from the earliest studies of the Sun was any close examination of the chromosphere. The neglect stemmed from a simple case of invisibility: In comparison with the bright photosphere below it, the chromosphere emits a relatively puny amount of light. This makes it visible to the naked eye at select times only—specifically, during a total solar eclipse, when the disk of the

Pith-helmeted astronomers from the Lick Observatory in San Jose, California, get ready for a total eclipse of the Sun in Jeur, India, on January 22, 1898. Trained skyward is the expeditionary party's forty-foot solar telescope, "Jumbo," which recorded almost every total solar eclipse worldwide between 1893 and 1931, when the development of the coronagraph permitted astronomers to study the Sun's outer atmosphere, or corona, in the absence of an eclipse.

Moon blocks out the photosphere and allows the chromosphere to emerge as a thin, bright red ring that encircles the lunar silhouette.

Modern astronomers need not await such an occasion, because they have a wide range of sophisticated solar detectors at their disposal. On Earth, for example, radio telescopes collect the chromosphere's emissions of radio waves, while in space an argosy of satellite-borne telescopes measure the chromosphere's output of ultraviolet light, x-rays, and gamma rays. (Absorbed by Earth's atmosphere, most such radiation never reaches the ground.)

This observational arsenal has enabled solar scientists to catalog the entire repertoire of chromospheric effects. These include evanescent streams of hot gas, called spicules, that rise 6,000 miles high into the chromosphere; although each spicule lasts only a few minutes, as many as 250,000 of them may be seen dancing above the Sun's surface at any moment. More impressive still are prominences, arcing loops of bright gas that often originate above regions of sunspot activity, penetrate the ceiling of the chromosphere, and reach far into the corona. Some, called quiescent prominences, form decorous arches that hold their shape for weeks or months before collapsing; others, known as eruptive prominences, explode from the chromosphere in a shower of gaseous shreds. Finally, bright, cloudlike structures called plages hover in the vicinity of sunspot groups; these bear the French name for "beaches" because they resemble stretches of bright sand against the chromosphere's darker background.

Nineteenth-century astronomers could hope to catch sight of such phenomena during a solar eclipse, but any such viewing was hard to come by. First, total eclipses of the Sun occur less than once a year; only eight will take place between 1990 and 2000, for example. Second, eclipses often are visible only from remote sites, such as the middle of an ocean or the Arctic tundra. Victorian astronomers therefore had to dismantle their delicate instruments and ship the pieces on rudimentary transportation systems to desolate locales, where they would be reassembled without technical support and operated without shelter from wind and weather. Finally, the observers had no guarantee that they would actually witness an eclipse: Rain, overcast skies, even the appearance of a few clouds at the wrong moment could easily ruin everything.

These obstacles drove solar scientists to seek easier ways of viewing the chromosphere and its showy displays. The first to find such a method was yet another amateur British skywatcher, Joseph Norman Lockyer, who had begun working as a clerk in London's War Office in 1857. The Crimean War had recently bloated the War Office staff, and many of the department's jobs were notorious sinecures; Lockyer's hours, for instance, were 10:00 to 4:00, with several of those hours allocated to a leisurely lunch and an afternoon walk through the park.

The comfortable regimen allowed Lockyer ample time to pursue his two hobbies, writing and astronomy. After joining the War Office, he even published a book called *The Rules of Golf*. But it was the stars, not literary stardom, that beckoned him most strongly.

DECODING THE SUN
Like many others in the field in the early 1860s, Lockyer sensed that the focus of astronomy was shifting from the infinite to the infinitesimal. This view arose mainly from a recent discovery by two German scientists, physicist Gustav Kirchhoff and his chemist research partner, Robert Bunsen. In 1859, the pair had shown that every chemical element, when heated to incandescence, emits a unique pattern of brightly colored lines at certain wavelengths of the spectrum of visible light. Kirchhoff further demonstrated that each element would absorb the same wavelengths from sunlight, producing dark absorption lines against a continuous bright emission background. Dark lines in the solar spectrum had been a mystery for nearly six decades, but now the mystery was solved: Certain wavelengths of light from the inner layers of the Sun were being absorbed by gases present in the Sun's atmosphere. Kirchhoff identified in the solar atmosphere at least six elements known on Earth, including sodium, calcium, and iron.

The primary tool of this elemental decoding was the spectroscope, an instrument that featured a prism mounted in front of a telescope lens. Until Joseph Lockyer's day, the spectroscope had been used to examine the light thrown off by the entire solar disk. Lockyer's scientific instincts, however, told him that the spectroscope could be tailored to study isolated features on the surface of the Sun. In 1866, he came up with a simple way to do this: He projected a telescopic image of the Sun onto a screen, then made a thin slit in the screen and channeled the light passing through the slit into his spectroscope. He could position the screen so that the slit admitted the light from a sunspot and the photosphere on either side and then examine the strength of the spectrum lines in both regions. Since the lines varied in strength with increasing temperature, Lockyer was able to prove that sunspots are cooler than the solar surface around them.

This success led Lockyer to wonder what his technique might disclose about other features of the Sun. "May not the spectroscope afford us evidence," he mused in a scholarly paper written in 1866, "of the existence of the 'red flames' which total eclipses have revealed to us in the sun's atmosphere, although they escape all other methods of observations at other times?" Lockyer theorized that these "red flames"—known today as prominences—consist of heated gas, which emits a distinctive spectral pattern of bright, clearly defined lines. On October 17, 1868, he began scanning the light coming from just above the limb, or edge, of the Sun in hopes that a prominence would betray itself through such a telltale emission.

On the third day after he inaugurated the technique, Lockyer peered steadily through a spectroscope in the garden of his Wimbledon home until he found the spectral signature he was looking for. "I saw a bright line flash into the field," Lockyer later recalled. "My eye was so fatigued at the time that I at first doubted its evidence, although, unconsciously, I exclaimed 'At last!' Leaving the telescope, I quitted the observatory to fetch my wife to endorse my observations." Lockyer thus became the first to witness a solar prominence in broad daylight. From the characteristics of the lines he observed, he concluded that prominences are composed primarily of hydrogen gas, with traces of other gases. At the end of a month of continuous viewing, he demonstrated that solar prominences belong to an extensive outer atmosphere, which he christened the chromosphere.

THE ENIGMA OF HELIUM

More intense scrutiny of prominence spectra led him to a second important discovery, one that would precipitate nearly thirty years of scientific debate about the Sun's constitution. When viewing light from various prominences through his spectroscope, Lockyer noticed a third yellow spectral line standing near the two that characterize the spectral emission of sodium. No element known at the time produced that particular pattern; he therefore concluded that the third spectral line belonged to some substance in the Sun that did not exist on Earth—or had not yet been identified. He named

George Ellery Hale, founder of the Mount Wilson Observatory near Pasadena, California, works the controls of a spectrograph located at the bottom of the observatory's sixty-foot tower telescope. In 1908, Hale's study of solar images captured by the spectrograph led him to conclude that sunspots were imbued with fierce magnetic fields.

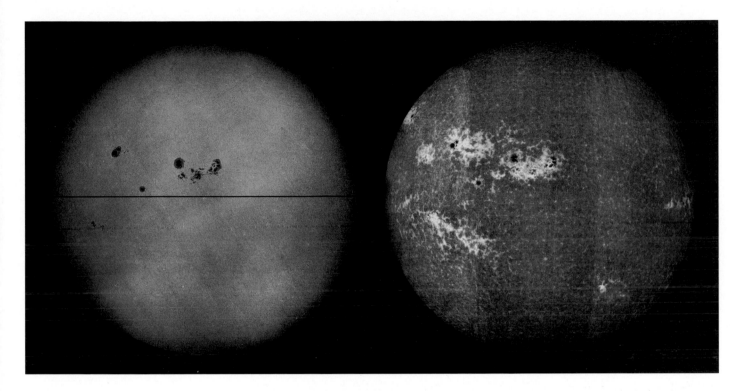

Side-by-side images of the solar disk taken in 1907 demonstrate the power of George Hale's invention, a modified spectrograph called the spectroheliograph. The above photograph, taken in ordinary visible light, shows a sunspot group in stark silhouette against an otherwise unblemished solar disk. The photograph at right, taken within a day of the first by a spectroheliograph admitting only the light given off by calcium atoms in a narrow band of the solar spectrum, reveals features such as plages, the bright patches that tend to hover above sunspots.

the enigmatic element "helium," after the Greek word for the Sun, *helios*.

Today, helium is used in a variety of ways, from inflating party balloons to cooling the circuits of supercomputers, but for nearly three decades after Lockyer's discovery, scientists failed to find it on Earth. This raised some troubling questions: If the universe contained elements that did not exist on the Sun's third planet, how could earthbound astronomers puzzle out their behavior? And how, assuming that elements like helium remained unattainable for laboratory analysis, could science ever divine the true nature of the stars? As it happened, a prominent British scientist, Sir William Ramsay, made such queries moot just before the turn of the century. In 1895, while analyzing a gas given off by a uranium compound, Ramsay realized that its spectral signature matched that of Lockyer's solar helium. He quickly communicated his finding to Lockyer, who confirmed that the gas was indeed the element he had posited.

Lockyer's success in modifying and refining the spectroscope to study the Sun presaged further inventiveness among astronomers. Back in the 1840s, an amateur American astronomer named John William Draper had fitted a spectroscope with a camera, thereby creating the spectrograph, and used the device to photograph the solar spectrum; by the 1870s, photography and astronomy had made such advances that Draper's son Henry succeeded in photographing the spectra of more than eighty stars. A few years later, an American telescope designer, George Ellery Hale, would customize the spectrograph and use it to make a major breakthrough in solar astronomy.

Encouraged by his father William, a wealthy manufacturer of hydraulic elevators in Chicago, Hale began building astronomical equipment in his teens. "My father always bought for me any books that I needed," he later recalled, "but in the case of instruments his policy always was to induce me to construct my own first and then to give me a good instrument if my early experiments were successful." Rising to this challenge as a fourteen-year-old in 1882, Hale constructed a small telescope for himself, whereupon his father purchased "an excellent four-inch Clark." Later, as a twenty-year-old physics major at the Massachusetts Institute of Technology, Hale drew up the designs for a laboratory equipped with a spectrograph, featuring a telescope with a ten-foot focal length, quite powerful for the time. William Hale then followed his son's blueprints to build the facility in the backyard of his Chicago home. The laboratory ultimately grew into Chicago's Kenwood Observatory, which remained in operation until the turn of the century.

A BOLT FROM THE BLUE

The idea for Hale's upgraded spectrograph—an instrument he called the spectroheliograph—simply came to him as he rode a Chicago trolley car one day in the summer of 1889. The device, completed that autumn, represented a tremendous advance over its predecessors. In addition to a slit to admit sunlight, Hale's spectroheliograph featured a second narrow slit, parallel to and behind the first one, that allowed only one spectral line at a time to enter the bowels of the instrument. The spectroheliograph thus functioned like a highly selective filter, permitting astronomers to view the Sun or a portion of the Sun in the wavelength peculiar to a certain element (the red of hydrogen, say, or the yellow of sodium) while blocking out the wavelengths emitted by all other elements in the Sun's atmosphere.

To create a composite photograph of the solar disk in any particular spectral line, Hale mounted his spectroheliograph on ball bearings. A motor-driven pulley then moved the instrument laterally in a slow and continuous fashion, enabling it to photograph the Sun in isolated strips that gradually accumulated across the photographic plate.

In this way, Hale captured detailed photographs of the Sun's surface features—notably, prominences and plages—in the light from several selected spectral lines. The device delivered its most striking images in the ghostly violet light emitted by calcium vapor. "Yesterday I got a prominence photo good enough to prove the success of the method," Hale wrote in a letter to his college friend Harry Goodwin in 1891, "and the result is that I am just now feeling pretty neat." Over the next forty-eight years, Hale's fascination with the Sun, which he called the "typical star," would spur him to establish three observatories: one for the University of Chicago in Williams Bay, Wisconsin; one atop Mount Wilson near Los Angeles; and one on Palomar Mountain near San Diego.

Observing the Sun from atop Mount Wilson in 1905, Hale detected some odd behavior on the part of the dark absorption lines that show up in the spectra

In 1908, while analyzing a spectrogram taken along a band bisecting a sunspot like the one shown below, George Hale noticed that the narrow lines created by light absorption on either side of the spot split into three separate lines inside the spot itself *(right)*. Hale recognized the pattern as an instance of the Zeeman effect (after its discoverer, Dutch physicist Pieter Zeeman)—the splintering of spectral lines when an element is subjected to strong magnetic forces. Hale's evidence for solar magnetism prompted Zeeman himself to congratulate Hale for having explained the nature of sunspots "in so satisfactory a manner."

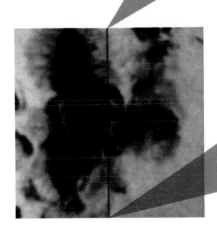

of sunspots: Many of the lines that normally appeared as solid bars in light emitted by unblemished portions of the Sun had instead split into several thin, closely spaced lines. Hale recognized the spectral division as an instance of the so-called Zeeman effect, named for the Dutch physicist who, in 1896, had discovered that light emitted by an element under the influence of a strong magnetic field displays fractured spectral lines *(above)*. The detection of the Zeeman effect in light streaming from sunspots suggested to Hale that magnetic fields must exist inside those spots.

Using a newly erected, thirty-foot-high vertical spectrograph at the Mount Wilson Observatory in June of 1908, Hale captured the first photographs of the Zeeman effect as it manifests itself in a sunspot spectrum. By comparing these images with the results of laboratory experiments in which he had measured the degree of spectral splitting caused by magnetic fields as high as 15,000 gauss, Hale proved that powerful magnetic fields do indeed govern the structure and behavior of sunspots. Hale also calculated that the strength of these fields went as high as 4,500 gauss, which solar scientists now deem to be several thousand times more powerful than either the Sun's or the Earth's overall magnetic field.

Although the precise makeup of sunspots continues to defy analysis today, Hale had made it clear that the spots were somehow linked with the fierce magnetic fields at play on the solar surface—or perhaps below. In a congratulatory letter to Hale in 1908, the president of the Carnegie Institution declared, "This is surely the greatest advance that has been made since Galileo's discovery of those blemishes on the sun."

Hale's contemporaries were aware that pairs of sunspots travel in groups, but in 1914 he refined their understanding by showing that the leading spot in any group always displays the reverse polarity of the spot at the rear. The group's leading and following spots thus behave like the north and south poles of a horseshoe magnet whose main body lies buried in the Sun. At any given time during the eleven-year sunspot cycle, the leading spot of a group in the northern hemisphere will be of one polarity, and the trailing spot will be of the opposite. In the southern hemisphere, meanwhile, the leading and trailing sunspots in each group will also exhibit opposite polarities, but their polarities will be a reverse image of those shown by the northern-hemisphere groups. Thus, if the leading spot in the northern hemisphere is a north pole, the leading spot in the southern hemisphere will be a south pole.

Further observations led Hale to realize that these polarities reverse from one sunspot cycle to the next: If the leading spots in the northern hemisphere are north poles at one time, then eleven years later they will be south poles. The true solar cycle, Hale thus established, lasts not eleven years but twenty-two: The Sun requires twenty-two years to resume its original state in terms of both sunspot activity and the orientation of its magnetic fields.

By the early 1960s, astronomers building on Hale's pioneering work had formulated a rough notion of how sunspots might arise and expire on the solar surface. Even as modified today, this theory still falls short of explaining all aspects of the cycle, yet most solar scientists agree that sunspots emerge through the interplay of the Sun's magnetic field, large-scale convection currents that bring hot gases to the surface from the interior, and differential rotation—that is, the Sun's tendency to revolve faster at the equator than at the poles. Because the Sun's magnetic field lines run beneath the surface from pole to pole, differential rotation gradually coils these field lines around the body of the Sun; eventually they become so twisted and stretched that they rise to the surface, forming sunspots where they break through. Apparently these eruptions have a cathartic effect on the Sun's overall magnetic field: As the twenty-two-year cycle continues, the sunspots somehow cause the field lines to gradually resume their orderly north-south orientation.

MYSTERIES OF THE OUTER LAYER
Through the diligence of such workers as Carrington, Lockyer, and Hale, the photosphere and chromosphere had begun to divulge their secrets. But the outermost layer of the Sun's atmosphere, the solar corona, remained chimer-

American solar scientist Charles Young prepares to study the total eclipse of May 28, 1900, through a prism spectroscope in Wadesboro, North Carolina. Using a similar device in 1869, Young had observed an unidentified spectral line of deep green emanating from the Sun's corona. Convinced he had discovered a new element, Young coined the term "coronium." The true nature of the line, however, would elude astronomers for another seventy years.

ical in the extreme. By 1878, astronomers had determined that the corona—visible from Earth during a solar eclipse as a glow of white, pearly light wreathing the dark disk of the Moon—changes its shape in time with the sunspot cycle. At sunspot maximum, many long white strands called coronal streamers radiate outward from the Sun like the petals of an enormous dahlia, while at sunspot minimum the corona displays a smooth edge with just a slight bulge at the solar equator.

The first clues about the composition of the corona came from an American astronomer named Charles Young. Possessor of a keen mind—he graduated first in his Dartmouth class at the age of eighteen—Young appeared destined to distinguish himself in astronomy. His grandfather, Ebenezer, and his father, Ira, successively held the Appleton professorship of natural philosophy and astronomy at Dartmouth from 1810 until 1858. Like Carrington before him, however, Charles Young initially embarked on a religious life; at twenty-one he entered the Andover Theological Seminary with plans of becoming a missionary.

In 1856, however, Young changed his mind, and within a decade he had revived the family tradition by assuming the Appleton chair. Young's new post entailed a good deal of field research, and on August 7, 1869, he found himself preparing to observe a solar eclipse in Burlington, Iowa. As the Moon bit into the limb of the Sun, Young saw a bright green line materialize in his spectroscope. But he was momentarily distracted. "I looked over my shoulder for an instant, and beheld the most beautiful and impressive spectacle upon which my eyes have ever rested. It could not have been for five seconds, but the effect was so overwhelming as to drive away all certain recollection of what I had just seen. What I have recorded I recall from my notes taken down by my assistant."

Despite Young's blurred memory of the event, the significance of the green line was crystal clear: Like the yellow line that Lockyer had observed less than ten months earlier, this green spectral line belonged to no element known on Earth. Astronomers of the day therefore concluded that the Sun's corona was made from some strange, possibly extraterrestrial substance. They labeled it coronium.

ENTER, THE CORONAGRAPH

There the mystery of coronal composition would rest for more than seventy years. In the meantime, a French scientist by the name of Bernard Lyot devised an instrument that permitted sungazers to make routine photographs of the solar corona.

As an apprentice astronomer at the Observatory of Meudon in 1920, Lyot had become intrigued with detecting the polarization—that is, the magnetic orientation—of light reflected from other planets in the Solar System. He hoped that by comparing the polarization measured in the light from a planet with that of the light scattered by substances on Earth, he could determine the composition of the other planet's surface. Unfortunately, the instruments of his day were unequal to the task, so in 1923 Lyot, undeterred, built his own polariscope. Ten times more sensitive than existing detectors, it enabled him to measure the polarization that resulted from the scattering of sunlight by particles in the Martian atmosphere. Lyot's research showed these particles to be sand, from which evidence he concluded—correctly, as the Mariner space probes revealed during the 1970s—that sandstorms rage across the surface of Mars.

Lyot was thwarted, however, when he turned his polariscope to the Sun. The polarization of its light could be measured only during an eclipse, which mutes the intense brightness of the photospheric background, yet the few moments afforded by such an event were insufficient to complete the readings. Lyot solved this problem in 1930 by designing a second device, the coronagraph, which would allow him to gauge the polarization of coronal light without waiting for an eclipse. In effect, Lyot's coronagraph created its own eclipse, blocking out the disk of the Sun in a specially built telescope so that the light streaming from the corona could be fed into a polariscope for analysis.

Masking the Sun was a fairly straightforward task, but Lyot faced another, more daunting challenge: He would somehow have to mute the brightness of the background light that is scattered by particles of dust in Earth's atmosphere. Although this scattered light disappears during an eclipse, blocking the solar disk in a telescope does not suffice to damp it. Lyot therefore had to devise a way to minimize this light in his coronagraph—which he did by constructing a special filter that was made up of stacked crystal laminae, or thin plates, with differing interference properties. The individual laminae could be layered in various ways to intercept the maximum amount of background light.

Lyot's instrument served to refocus scientific attention on the solar corona. Indeed, within a decade of its invention, another scientist—Swedish astrophysicist Bengt Edlen—would solve, once and for all, the riddle of Charles Young's coronium. That mysterious green line, Edlen proved in 1941, matched the spectral line emitted by an iron atom stripped of thirteen of its normal complement of twenty-six electrons.

Nevertheless, the green line continued to baffle members of the astronomical community. They calculated that fantastically high temperatures—one million degrees Kelvin, at a minimum—would be required to keep iron atoms in such a denuded (or ionized) state, yet that extreme level of coronal heating contradicted all existing suppositions about the corona. Because the corona lies farthest from the surface of the Sun, solar scientists had been maintaining since the nineteenth century that it must be the coolest atmospheric layer of all.

Confirmation that searing conditions do indeed prevail in the corona came in 1946, when the advent of radio astronomy permitted solar physicists to gauge the inconceivably hot temperature of the corona on the basis of

Photographed in the light of highly ionized iron, the Sun's corona appears as a fountain of verdant light. In 1941, Swedish physicist Bengt Edlen determined that this emerald glow was emitted not by Charles Young's coronium but by iron whose atoms had been stripped of half of their twenty-six electrons. Edlen's discovery caused solar scientists to raise their estimates of the corona's temperature to millions of degrees Kelvin—hundreds of times hotter than they had previously thought.

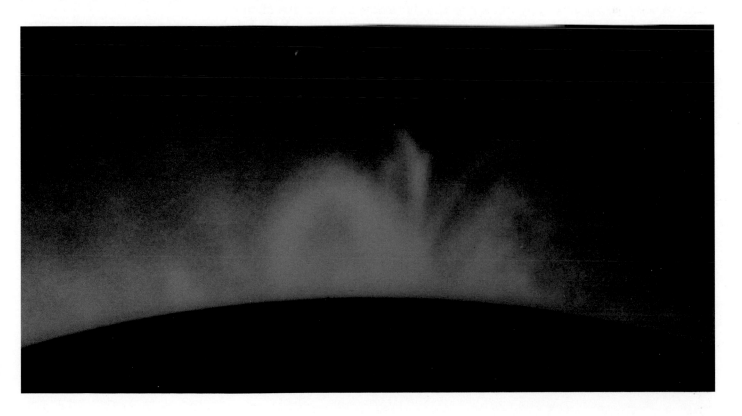

the radio wavelengths it emitted. Then, in the 1950s, continued spectral observations revealed the presence of other highly ionized atoms. Not only were nickel, calcium, and argon ions swirling through the corona, but iron atoms lacking as many as fifteen electrons turned up as well. These findings pushed estimates of the corona's temperature higher still.

Having established that the corona is hotter than any part of the Sun but its core, solar scientists set out to discover why. Their investigations, which continue to this day, have uncovered two possible mechanisms so far. The first involves the solar equivalent of sonic booms. As hot gases well up through columns in the Sun's interior layers, they create pressure waves that reach the solar surface. The majority of these waves are refracted back from the surface toward the core and remain trapped in the Sun, but a small percentage of them manage to escape the photosphere and propagate outward through the corona, heating the coronal gases as they go. The second and more likely possible source of coronal heating is the dissipation—and ensuing conversion to heat—of magnetic waves traveling outward along the magnetic field lines that stretch from beneath the Sun's surface into the corona and beyond.

The search for the engine that generates the corona's high heat signaled a subtle shift in solar physics, one that had started around the same time Lyot was perfecting his coronagraph. In addition to studying the Sun's surface effects, solar scientists now began to delve deeper. The object of their quest: to piece together the very functioning of the thermonuclear furnace at the heart of the Solar System.

SOLAR ANATOMY

L ike a giant physics laboratory in the sky, the Sun offers scientists a chance to study a star at close range. Using sophisticated telescopes and computers, these researchers have devised a number of plausible models to explain the Sun's nature and dynamics, illustrated on the following pages. Many of the most significant breakthroughs have come in the last three decades as astronomers refined their understanding of well-known surface features such as sunspots and began to discern the mechanics of more exotic, less easily observed phenomena such as coronal holes. They even developed a new science—helioseismology, or the study of the oscillating solar surface—that promises to open a window on the star's hidden core.

Deep in that core hums a mighty thermonuclear furnace, whose tremendous temperatures and pressures force the nuclei of hydrogen atoms to fuse into helium, creating photons of energy in the process. Streams of charged electrons, shaken loose by the collisions of superheated matter deep in the Sun, align themselves along powerful lines of force that in some areas concentrate into magnetic fields 5,000 times stronger than the one surrounding Earth. Bending and braiding as the Sun turns, the field lines create spots, streaks, and geysers on a solar surface in continual turmoil.

Finally, one of the newest areas of study is the Sun's atmosphere, three vaguely defined layers of gas called the photosphere, the chromosphere, and the corona. The topmost tier, the corona, extends great distances into interplanetary space, where it shimmers and blazes at temperatures that continue to defy theoretical predictions.

Bursting from the Sun at the speed of light, the subatomic particles called neutrinos bombard every square centimeter of Earth at the rate of 70 billion per second. Passing unimpeded through almost everything they encounter, neutrinos produce few observable effects—but their numbers hint at the prodigious amounts of energy unleashed by the Sun.

SOLAR ALCHEMY

The proton-proton chain. The conversion of hydrogen into helium begins when two protons *(red)* fuse into a deuteron composed of a neutron *(white)* and a proton, releasing a positron *(black)* and a low-energy neutrino *(green)*. The deuteron and another proton then fuse into helium-3 (two protons plus one neutron), giving off energy in the form of a gamma ray *(purple)*. Finally, two helium-3 atoms fuse, forming an atom of helium-4 (two protons plus two neutrons) and two protons.

The beryllium-boron reaction. In a rare branch of the proton-proton chain, atoms of helium-4 and helium-3 merge into a nucleus of beryllium-7 (four protons plus three neutrons), releasing gamma ray energy. A proton then transforms the beryllium-7 into boron-8 (five protons plus three neutrons), which emits a second gamma ray before it decays into a positron, a high-energy neutrino, and beryllium-8 (four protons plus four neutrons). Finally, the beryllium-8 splits into two helium-4 nuclei.

STOKING THE FURNACE

Each second, the Sun transforms 700 million tons of hydrogen gas into 695 million tons of helium gas through the nuclear reactions illustrated below. The remaining 5 million tons of matter (about 600 times the weight of water flowing over Niagara Falls in one second) escape as pure energy.

The thermonuclear powerhouse is contained—and maintained—by the extreme conditions that exist at the Sun's core: temperatures of 15 million degrees Kelvin and pressures about 250 billion times that of Earth's atmosphere. Since the core holds approximately 50 percent of the Sun's mass in just one sixty-fourth of its volume, the Sun is saved from inward gravitational collapse only by the outward pressure

exerted by the stupendous heat generated in the core.

Although solar astronomers have learned much about the Sun's energy production, one puzzle is the subject of particular focus. Scientists have determined that the conversion of hydrogen to helium yields byproducts known as neutrinos—particles of no electric charge and virtually no mass that travel out in all directions and pass through nearly everything in their path. Yet elaborate devices designed to detect them have found only one-third the number of particles predicted by the prevailing theory of solar physics. Despite a multitude of possible reasons for this so-called neutrino deficit, the case of the missing neutrinos remains open.

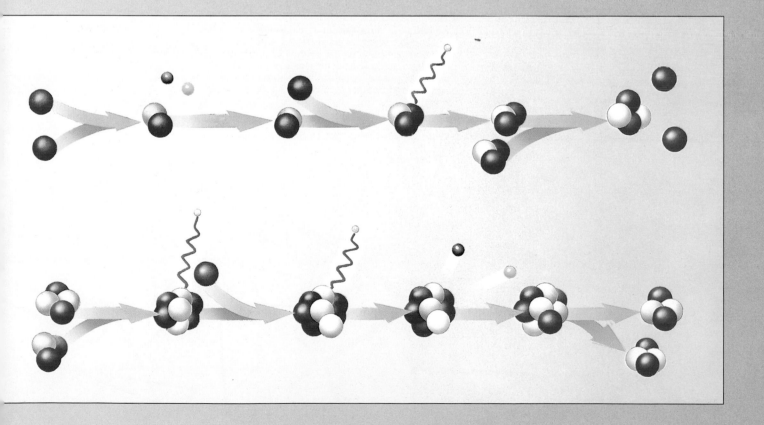

Released at the Sun's core in the form of high-energy, short-wavelength gamma rays, photons trace jagged paths through a region known as the radiative zone, randomly changing direction each time they strike an atom of gas. With each collision, the photons are robbed of a tiny bit of energy; as their wavelengths elongate, the solar radiation downshifts immediately into X-rays, then ultimately into the still-longer waves of visible light.

A Photon's Progress

The same fusion process that spawns neutrinos at the core of the Sun also releases tiny packets of solar energy known as photons, which travel a circuitous path to the solar surface, or photosphere. Theirs is an odyssey that occurs in two parts, reflecting the different ways by which the energy is transported through the Sun's inner layers.

In the first stage of the journey, a photon generated at the core moves outward into the so-called radiative zone, a region of densely packed atoms of hydrogen and helium gas where energy is transferred by the process of radiation. Here the photon's progress is unpredictable: Over and over again, it slams into atoms that absorb the photon and then emit it on a divergent course. The photon loses a tiny fraction of its energy with each collision, but it continues to meander toward the surface by heading for regions of lower density and temperature.

After many thousands of years of following the path of least resistance, the photon reaches a level, about 130,000 miles below the surface, where the solar gases have cooled to approximately two million degrees Kelvin. This critical threshold marks the lower boundary of the convection zone, a region whose reduced temperatures transform the Sun's gases into insulators against radiation; in place of the radiative process, turbulent convection currents carry solar energy to the Sun's surface.

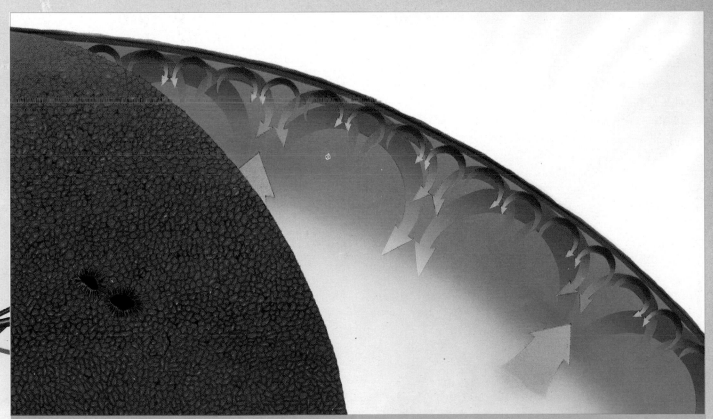

The granulated pattern of the photosphere—the Sun's visible surface—is created by alternating flows of hot and cool gas just beneath the surface. The granules form and disappear in a regular rhythm of about eight minutes. The dark blotches are sunspots, each one about the size of Earth; they appear dark only because they are cooler than the surrounding photosphere.

In the convection zone, roiling currents of hot gas rise, expanding and dispersing much of their heat at varying levels below the Sun's surface. The cooled gases then sink, only to become reheated and rise again.

Where Magnetism Reigns

Like Earth and many other planets in the Solar System, the Sun possesses a magnetic field. But unlike its planetary counterparts, whose lines of magnetic force tend to run tidily from one pole to the other, the Sun's magnetic field can be notoriously unruly. As shown at right, field lines that begin as orderly meridians gradually become twisted beyond all recognition. After about eleven years, however, the tangles unravel and the lines straighten themselves out. Because the incidence of sunspots, flares, and other solar outbursts rises as the field lines become more entwined, this eleven-year period is often referred to as the solar activity cycle.

Two powerful forces act in concert to contort the field lines. One is differential rotation. Because the Sun is gaseous, not solid, its various regions rotate at differing speeds. Areas near the solar equator, for example, complete a rotation every twenty-seven days, whereas those near the Sun's poles require thirty-four. Because the field lines are "frozen" into the gases by electrical conductivity, the lines are dragged along as the gases move at their uneven rates. The speed differential pulls the equatorial regions of the field lines faster and farther around the Sun, eventually coiling the lines about the solar globe.

As the once-vertical field lines wrap horizontally around the star, they become concentrated closer together and form magnetic tunnels, or flux tubes, that measure about 300 miles in diameter just below the Sun's surface. A second powerful force—convection— also comes into play: Bubbles of hot solar gases, rising through the Sun's convection zone, wash over the flux tubes, intertwining them like magnetic braids. This raises the pressure inside the flux tubes, expelling the gases they contain and making the tubes lighter than their surroundings. The flux tubes then float upward, forming a pair of sunspots where they break through the surface.

There is a limit to this process of magnetic knotting. In fact, it is fully reversible: At the peak of the solar activity cycle, when the flux tubes are most densely concentrated, corkscrew motions of the solar gases impel nearby field lines to merge and recombine into vertical lengths. This reconfiguration continues until it has untangled the Sun's magnetic skein.

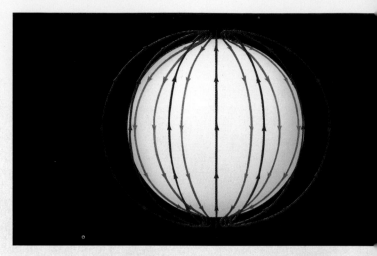

At the beginning of the eleven-year cycle of solar activity, the Sun's magnetic field lines run directly from one pole to the other, and the number of sunspots and other surface blemishes is at a minimum. A compass on the Sun would point to the star's north pole.

As convection currents beneath the solar surface augment the distorting effects of differential rotation, the field lines become so twisted and braided that they begin to kink up. Bundled field lines known as flux tubes pop through the surface at relatively high latitudes, where the shearing effect is greatest, giving rise to sunspots.

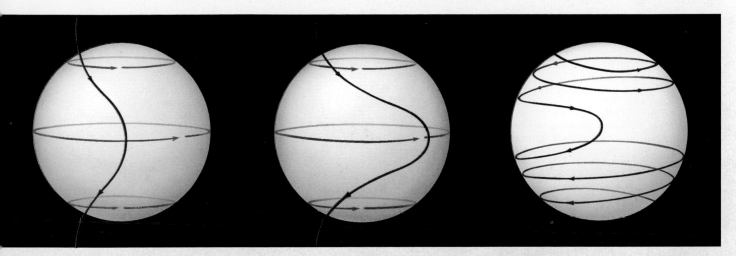

Under the influence of differential rotation—the Sun's tendency to rotate faster at the equator than at the poles— each magnetic field line is tugged out of shape and begins to lose its vertical orientation. (For clarity in the time-lapse sequence that begins here, a single magnetic field line serves to represent the Sun's multitude of magnetic field lines.)

Just a few months into the solar activity cycle, the Sun's differential rotation has yanked the field lines so far out of position that their midsections extend as much as 120 degrees of longitude around the solar globe from their trailing ends.

The lines wrap around and around the Sun, becoming so stretched that they assume a horizontal orientation. The field lines in the northern hemisphere have an opposite magnetic direction from those in the southern hemisphere.

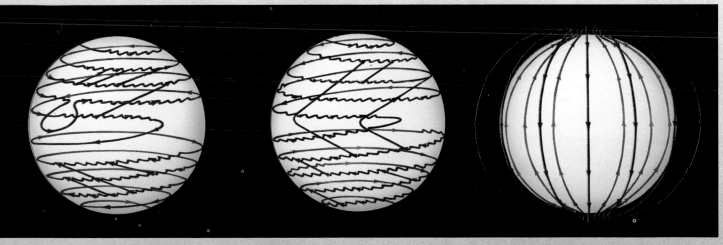

At the peak of the solar activity cycle, the magnetic field lines are extremely tangled and looped, and sunspots erupt closer to the equator. Cyclonic motions in the solar gases, however, help drive together field lines of like magnetic orientation; this magnetic reconfiguration, as it is called, will eventually bring the cycle of magnetic turmoil to a momentary end.

As the cycle nears an end, the field lines rejoin in a more stable configuration; in response, the Sun's overall magnetic field becomes more evenly distributed and the number of sunspots decreases. A new direction for the magnetic field, opposite the one at the cycle's start, begins to emerge.

At cycle's end, sunspots and other solar disturbances subside, and the field lines revert to their longitudinal symmetry. The polarity of the Sun's magnetic field is reversed from its orientation eleven years before: A compass on the star would now point toward the south pole.

Blanketing the pebbly surface of the Sun, the chromosphere *(pink)* stages a host of dynamic solar features. Plages (from the French word for "beaches") are bright patches *(below, far left)* that hover just above active areas such as sunspots on the surface. Their luminosity may result from the abnormally high flux, or intensity, of their magnetic fields. Flares *(far right)* are violent releases of pent-up solar energy that spew gaseous matter into the corona and beyond. Filaments *(upper right)* are arches of suspended gas, some of them more than 400,000 miles long and 50,000 miles high, thought to form the boundary between regions of opposite magnetic polarity. Although they appear dark against the disk of the Sun, filaments loom as bright prominences when silhouetted against the blackness of space.

INTO THE SPHERE OF COLOR

Observers of a total solar eclipse are likely to witness a tantalizing sight just before and after totality: A fiery red ring blazes into view around the disk of the Sun, lingering for a period of several seconds and then suddenly vanishing. This fleeting apparition is the lower atmosphere of the Sun, known as the chromosphere. Ordinarily obliterated by the intense light of the photosphere, the chromosphere reveals itself to be an exotic preserve when studied through special telescopes and highly discriminating filters. The tenuous region is home to all manner of undulating features that recall some of Earth's own natural forms: a hedgerow, say, or a patch of burning prairie grass, or a cluster of mountain peaks—all of them backlit by a vibrant crimson glow.

Engendering this diversity of forms is magnetism, the dominant force throughout the solar atmosphere. The Sun's ionized gases—collectively referred to as plasma—are far less concentrated in the chromosphere than they are in the body of the Sun itself; thus, unlike the dense gases in the Sun's lower regions, plasma in the chromosphere is unable to contain the Sun's magnetic fields. Instead, the magnetic fields dictate the plasma's behavior, sometimes keeping it at a slow simmer, at other times detonating it in a spectacular display of solar fireworks.

Fencelike rows of spicules—near-vertical cylinders of chromospheric gas that may weigh one million tons each—stand sentinel around supergranular cells some 20,000 miles in diameter. The long, curved strands of gas below are called fibrils; like plages, they indicate an active region nearby.

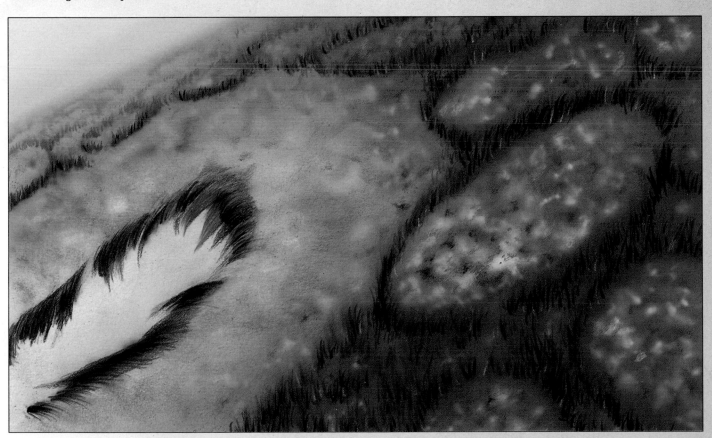

THE BLISTERING OUTER LIMITS

From the roof of the chromosphere, the Sun's corona—a realm of superheated gases so tenuous that 1.2 million cubic miles of it would weigh only one pound—reaches millions of miles into space. It is most familiar as the luminous white halo that surrounds the solar disk during a total eclipse, an effect produced by photospheric light bouncing off free electrons in the corona. As shown below, the corona also provides the backdrop for the Sun's most impressive light show of all: the arcing, pulsating patterns of incandescent gas known as solar prominences.

To astrophysicists, the corona is part chameleon, part conundrum. It changes shape in synchronization with the solar activity cycle, forming a jagged ring around the Sun at the peak of the cycle and then settling down as the sunspots and other surface events recede. By cycle's end *(opposite)* it is much transformed, trailing wispy plumes and streamers five million miles and more into the Solar System.

But the aspect of the corona that has baffled scientists for decades is its temperature. Contrary to the laws of thermodynamics, which hold that heat cannot be radiated from a cooler body to a warmer one, the temperature of the Sun's atmospheric layers climbs steadily from a minimum of 4,200 degrees Kelvin in the chromosphere to more than a million degrees Kelvin in the corona. Most astronomers today attribute coronal heating to magnetic waves transported along the same magnetic field lines that sculpt the corona's features. As with so many other solar processes, however, the details of this energy transfer remain to be worked out.

Loop. A graceful loop prominence rises above an active region of the solar surface, dramatic evidence of the magnetic fields that pervade the Sun's corona. The prominence took shape when hot plasma, captured by magnetic field lines, condensed and began to rain back into the chromosphere.

Quiescent. This type of monumental solar prominence—standing about four Earth diameters tall—is dubbed "quiescent" because it remains relatively stable until disappearing. Cool plasma, insulated from the searing heat of the corona by powerful magnetic field lines, molded itself into the shape of a hedgerow above an inactive surface region. Here the quiescent prominence may linger for months.

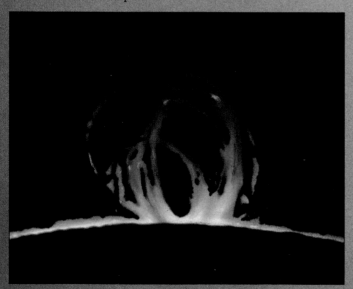

Eruptive. As shown below, shifts in the undergirding magnetic field of a quiescent prominence can cause the structure to explode without warning. As the eruptive prominence expands outward through the corona, its magnetic field lines are pulled asunder; finally the arch bursts *(far right)*, and its gaseous detritus either cascades back toward the surface of the Sun or escapes with the solar wind.

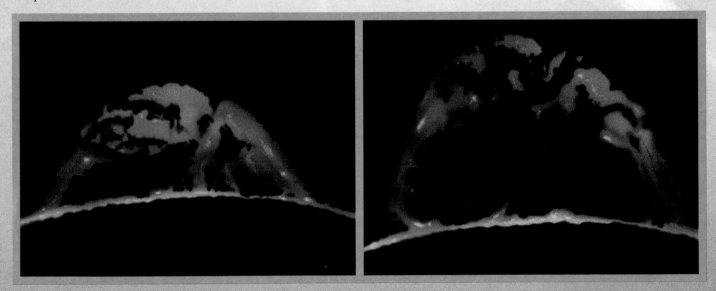

Deep in South Dakota's Homestake gold mine, beneath 4,850 feet of solid rock, a 100,000-gallon tank of chemicals lies in wait for the elusive solar particles known as neutrinos.

early a mile below the surface of the Earth, in a dimly lit cavern of the Homestake gold mine near Lead, South Dakota, a huge cylindrical tank lies on its side. It contains 100,000 gallons of the solvent per-chloroethylene. Known more simply as "perc," the chemical solution is used primarily in the dry-cleaning trade to remove stains from fine woolens and other garments. Though the Homestake mine, established in 1876, still yields nearly 400,000 troy ounces of gold a year, the tank of solvent has nothing to do with mining operations. It just sits, a trap awaiting its quarry.

The surface of the Earth directly above the tank, like every other spot on the globe, is constantly bombarded by everything from ordinary sunlight to tiny bits of cosmic dust that originate deep in space. However, all but one of these intruders are thought unable to lance through the 4,850 feet of rock above the tank. The single known exception is a subatomic particle called a neutrino, a by-product of nuclear reactions commonly accepted as the source of the Sun's heat. Because a neutrino lacks an electric charge and is either massless or almost so, it rarely interacts with other particles, passing through Earth almost as if the planet were not there. Indeed, intercepting these neutrinos on their way to Earth would require a shell of lead around the planet three light-years thick.

Almost all the neutrinos that strike the tank of perc zip right through, but occasionally a neutrino collides with one of the four chlorine atoms contained in every one of the tank's approximately 250 thousand quadrillion (250 followed by eighteen zeros) molecules of cleaning fluid. Instantly, the chlorine atom is transformed into a radioactive form of argon, which is culled from the perc by helium that is bubbled through the tank from time to time. Separated from the helium, the argon atoms are then counted as a measure of the number of solar neutrinos striking Earth.

That scientists would commission the excavation of 7,000 tons of rock to form a chamber for a subterranean tank having the volume of an Olympic swimming pool, then baby-sit the apparatus as it captured an odd neutrino or two from the Sun, bespeaks the acuteness of their hunger for information—any information at all—about the interior of this star. Direct observation of the Sun's depths is of course impossible because of the intense heat and, barring future technologies that would seem far-fetched in a science-fiction novel, will remain so. For hundreds of years, astronomers believed that their inability to probe beneath the fiery surface of the Sun and its stellar cousins was a permanent limitation. Proclaimed nineteenth-century French philosopher Auguste Comte about stars: "We can never learn their internal constitution."

Comte died in 1857, just two years before German scientists Gustav Kirchhoff and Robert Bunsen proved through spectrographic study that the Sun's atmosphere is composed chiefly of hydrogen. And within a decade of the philosopher's death, German physicist Hermann L. F. von Helmholtz and the British scientist William Thomson, Lord Kelvin, had worked out a seemingly plausible explanation for why the Sun is hot. Gravitational attraction between solar gas molecules, they said, caused the Sun to contract, compressing the gases and heating them to incandescence. Calculations indicated that the reduction in solar diameter would amount to no more than a few score feet each year. At that rate, the Sun would shine for many millions of years.

The work of Helmholtz and Kelvin was refined in 1870 by Jonathan Homer Lane, an American physicist at the U.S. Patent Office, and again in 1907 by Swiss physics professor Robert Emden, producing the so-called Lane-Emden theory of the stars. According to the theory, gas molecules close to the surface of the Sun would be far enough apart to vibrate independently of one another. But deeper in the interior, pressure would rise and the molecules would be squeezed ever closer together. Near the center, where a thermometer would indicate a temperature of about 12 million degrees Kelvin and a barometer would register a pressure 123 billion times that of the air pressure at the surface of the Earth (a unit of measurement called an atmosphere), the molecules would touch, preventing further compression. Meanwhile, the Sun would give off its heat to the surrounding void, cooling slightly and shrinking, albeit imperceptibly. The shrinking would put additional pressure on the remaining gaseous molecules. Following this process to its logical conclusion led astronomers to believe that the Sun would gradually dwindle in size until all its molecules had compressed and cooled into a frigid lump.

UNSETTLING EVIDENCE FROM THE PAST
The Lane-Emden theory seemed to account for everything that was known about the Sun: its mass, calculated by measuring its gravitational influence on the Earth; its diameter; the amount of heat and other energy it emitted; and the simple observation that in all the history of humankind's sungazing, the star had not become noticeably larger or smaller, hotter or cooler. Indeed, calculations indicated that for the Sun to have its turn-of-the-century mass, it must have been born 22 million years earlier and had about 17 million years more to live.

Though the Lane-Emden hypothesis would turn out to be wrong about why the Sun glows, it was backed by a sound investigative approach. Comte and others who shared his view had examined the problem using an inappropriate metaphor. The task of astronomers is not literally to see to the center of the Sun but rather to interpret the messages that the Sun sends their way. The British astronomer Arthur Stanley Eddington, an avid reader of mystery novels, once likened the process to analyzing the clues in a crime. Trained in mathematics at Trinity College, Cambridge, and practical astronomy at the Royal Observatory in Greenwich and at Cambridge Observatory, Eddington believed in the Lane-Emden theory—as did most other prominent scientists interested by such matters. He noted, however, that all the hints about the nature of the Sun encompassed by this hypothesis had been acquired by focusing intently on the gaseous ball glowing eight light-minutes from Earth.

Eddington was inclined to look farther afield. Just as a police investigator expects the scene of the crime to yield only some of the evidence in a case, the detective-minded scientist was alert to the possibility that glimmers about the Sun might come from other sources. In particular, precursors to the carbon-dating process, the universally accepted method of establishing the age of antiquities by determining the amount of carbon-14 present in a

THE ROAD TO THE CORE

Ignoring the gloomy assertion of French philosopher Auguste Comte that the interior of stars could never be known, astronomers and physicists began in the late nineteenth century to probe the Sun. After many false steps, the emergence of Albert Einstein's theory of relativity, combined with advances in particle physics, led the six scientists pictured here to develop a clearer—if still imperfect—understanding of the processes that power the solar furnace.

In 1932, British physicist James Chadwick discovered the proton's counterpart, the neutron, a further step toward the eventual realization that nuclear fusion powers the Sun.

Pondering the implications of Einstein's relativity theory, the brilliant British astronomer Arthur Stanley Eddington proposed in 1926 that the Sun burns by converting mass into energy.

Austrian physicist Wolfgang Pauli hypothesized the existence of neutrinos in 1931. Firm evidence came in the 1950s, but his work suggested the possibility of detecting neutrinos from the Sun.

sample, had shown that some rock on Earth was at least a billion years old. Furthermore, fossils of primitive creatures, unthinkable without life-sustaining warmth from the Sun, had been found etched in rocks. An inescapable conclusion followed: The Sun had to be much older than the tens of millions of years allowed by the Lane-Emden theory. Clearly, a better explanation was required.

Eddington contributed a major part of the answer in his 1926 classic, *The Internal Constitution of the Stars.* There he argued that the only possible source of the Sun's heat and light must be the conversion of some of its mass into energy according to the most familiar expression of Einstein's theory of relativity: $E = mc^2$. Eddington believed that hydrogen was being transformed into helium, but the precise nature of this transformation eluded him—and for good reason. Describing it would require concepts of the subatomic world that either were incomplete or had not yet been thought of.

But progress was rapid. For example, quantum mechanics, which explains the behavior of particles involved in nuclear reactions, was proposed in the late 1920s by German physicists Werner Heisenberg and Erwin Schrödinger. One by one, the ingredients for such processes were identified. In 1919, Cambridge physicist Ernest Rutherford proved that hydrogen has a nucleus

American physicist Carl Anderson—seen here with his cloud chamber—in 1932 discovered the positron, a positively charged elementary particle that resides outside the atomic nucleus.

German-born physicist Hans Bethe *(left)* and American physicist Charles Critchfield *(right)* demonstrated in 1938 that hydrogen fuels most stars, with carbon the catalyst and helium the ash.

45

Physicists Frederick Reines *(near left)* and Clyde Cowan stand before a diagram showing the successful result of "the hardest physics experiment they could think of": finding the neutrino, a chargeless, possibly massless particle emitted by the Sun. The small and large blips on the hand-drawn oscilloscope represent gamma ray bursts triggered by an antineutrino during nuclear reactions inside a South Carolina nuclear power plant. The gap between bursts—5.5 microseconds—was almost precisely what Reines and Cowan had predicted, allowing them to confirm the neutrino's existence.

made up of a proton. Thirteen years later, in 1932, James Chadwick, then working at Cambridge, discovered the neutron, an uncharged component of atomic nuclei that has approximately the same mass as the positively charged proton. The same year, American physicist Harold Urey isolated deuterium, a form of hydrogen with a nucleus containing a neutron as well as a proton, and his compatriot Carl Anderson found the positron, an electron with a positive charge.

In the spring of 1938, many of the world's most accomplished physicists and astronomers gathered at a historic meeting in Washington, D.C., to discuss the question of the Sun's power source. Cornell University's Hans Albrecht Bethe and Charles L. Critchfield, a student of quantum mechanics from

George Washington University, put their heads together and came up with an answer: The Sun is powered by nuclear fusion. They suggested that more than 98 percent of its energy comes from the transformation of hydrogen into helium through a process called the proton-proton reaction. The three-step process begins as a collision between two hydrogen nuclei and ends with the formation of a helium nucleus weighing seven-tenths of one percent less than its protons and neutrons taken separately. This difference in mass is converted to energy. Most of the proton-proton reaction's energy takes the form of gamma rays—packets of short-wavelength electromagnetic radiation. The remainder is embodied in a neutrino.

At the time, this particle lived in theory only. It had been dreamed up by Austrian physicist Wolfgang Pauli in 1931 to solve a perplexing puzzle in nuclear physics. Some nuclear reactions appeared to violate the principle of conservation of energy. That is, the energy coming out of the reactions was less than the energy going into them. Experiments failed to turn up the missing energy. Pauli expressed the frustration of many of his physicist colleagues when he said upon contemplating the problem: "In any case, it is too difficult for men, and I wish I had been a movie comedian or something of the sort and had never heard of physics." Not long thereafter, he proposed an entirely new type of particle having no charge and no mass that traveled at or near the speed of light and carried off the missing energy in an undetectable manner. Physicist Enrico Fermi christened this hypothetical entity with an Italian word meaning "little neutral one."

FLASHES OF ANNIHILATION

In 1956, Frederick Reines and Clyde Cowan, two researchers at the Los Alamos Scientific Laboratory (renamed the Los Alamos National Laboratory in 1981), devised an experiment to catch some neutrinos. According to theory, every particle in the world of atomic physics has, by definition, an antimatter counterpart, and vice versa. The two scientists intended to find an antineutrino, which would be as good as detecting the neutrino itself. Their calculations indicated that by-products of a collision between an antineutrino and a hydrogen nucleus would produce three tiny flashes of light in a special detector; this unique signature would prevent the researchers from mistaking sparkles from other reactions as a sign of their quarry.

Reines and Cowan set an antineutrino trap. Named Project Poltergeist, it consisted of a tank containing 1,000 pounds of water sited next to a nuclear reactor at the U.S. Atomic Energy Commission's Savannah River plant in South Carolina. Engaged in making plutonium and tritium for use in nuclear weapons, the reactors sustained processes that resulted in vast numbers of antineutrinos—or so theory promised. Reines and Cowan submerged their detector in the water and positioned sensitive photomultiplier tubes to record the anticipated trios of light flashes. To their great satisfaction, the apparatus succeeded in detecting antineutrinos at the rate of two or three an hour.

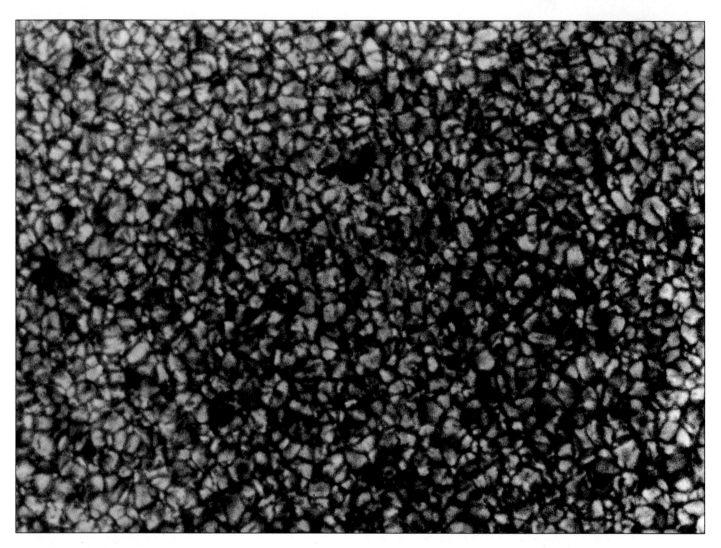

REASSESSING THE SUN

By the early 1960s, a new portrait of the Sun had emerged—a physical description based on Bethe and Critchfield's assertion that stars are fusion engines. Commonly known as the standard model, it differs radically from the picture drawn by the Lane-Emden theory. The standard model posits a spherical core surrounded by a gaseous region called the radiative zone, which is in turn encompassed by a layer known as the convection zone.

The core, about 275,000 miles across, represents nearly a third of the Sun's diameter. The temperature at the center is more than 15 million degrees Kelvin, and pressure is about 225 billion atmospheres. Under these conditions, atoms lose their identities and disassemble into plasma—a soup of protons, neutrons, and free electrons that zip around at speeds of 100 miles per second and more. Quintillions of times each quintillionth of a second, hydrogen protons slam into each other to form helium. The result is a solar

The granulated surface of the Sun, captured in a high-resolution photograph taken at France's Pic du Midi Observatory in 1977, is the product of tubular convection cells: The bright spots are areas where convection has delivered hot gases to the surface; the dark lanes separating the spots show where cooled gases descend toward the interior. Each granule lasts about ten minutes and is roughly 1,300 miles across—about half the width of the continental United States.

power output of some 380 sextillion kilowatts—the equivalent of converting, each second, five million tons of matter into energy. Were it not for the huge quantities of gas in the radiative and convection zones, the Sun would long ago have blown up, like a hydrogen bomb. The mass of gas exerts an inward pressure on the core that balances its explosiveness.

Although the Sun might seem to be consuming itself at a prodigal rate, the loss of material is actually insignificant. Given the amount of energy the Sun radiates, and its present mass and size, astrophysicists calculate that it has consumed only a few hundredths of one percent of its original mass. Moreover, the standard model places the birth of this medium-size star more than four billion years earlier than the Lane-Emden theory, a date consistent with the fossil and rock-dating evidence that had so upset the astrophysics apple cart.

More significant than the Sun's conversion of mass to energy is the depletion of hydrogen nuclei. According to the standard model, about half of the hydrogen mass originally in the core—the only region where fusion can occur—has already been converted into helium. Because heat and pressure in the core preclude the entry of additional hydrogen from outside, the Sun is approaching middle age. About five billion years from now, after consuming all its fusionable hydrogen, it will become a type of star called a red giant. In the process, it will expand to 100 times its present diameter, possibly vaporizing the Earth. Then the Sun will die, eventually becoming colder than the dark side of the Moon.

EN ROUTE TO THE PHOTOSPHERE
The Sun's radiation originates in the core, where the standard model predicts not a blinding brilliance but a deep blackness; none of the energy produced there lies within the visible spectrum. About one-millionth of the total takes the form of neutrinos, which escape directly to the surface of the Sun at light-speed (the trip takes about two seconds) and continue into space. Gamma ray photons, the main constituent of the star's energy, are equally invisible. After traveling less than a centimeter, a gamma ray collides with one of the many electrons in the swarm of subatomic particles at the core, producing x-ray photons that also cannot be seen.

These x-rays carom randomly from electron to electron, following a tortuous path of least resistance. That is, they migrate inexorably in the direction of decreasing pressure, temperature, and density toward the Sun's surface. Because of this continual ricocheting, a photon makes excruciatingly slow progress outward through the core. Even though it travels at the speed of light, an x-ray originating at the center would take more than 26,000 years to arrive at the outer limit of the core, 137,000 miles from the center. There, temperature and pressure have declined to such an extent—about six and a half million degrees Kelvin and nine billion atmospheres—that no more protons collide to form helium, and fusion ceases.

X-rays passing through the radiative zone transit regions of progressively lower temperature and pressure, where the distance between electrons increases and the rate of collisions between x-rays and electrons falls. In this region of the Sun, some free-flying electrons are captured by helium nuclei to form helium ions (helium atoms without a full complement of electrons). X-rays that encounter the electrons of these incomplete atoms are absorbed and their energy reissued as x-rays of longer wavelength.

About 300,000 miles from the center of the Sun, the radiative zone meets the convection zone, which extends to the photosphere. Huge surges of ionized gases drift outward through the convection zone at speeds thought to be greater than 200 miles per hour. Breaking the surface, these eddies disperse and cool, and their gases return toward the center. X-rays passing through the convection layer are absorbed by ions and converted to photons in the ultraviolet and infrared regions of the spectrum. Only at the surface is radiation produced in the visible spectrum—sunlight.

Among the insights offered as physicists worked out the details of the standard model in the early 1960s was the quantity of neutrinos generated in the Sun's core, a truly astronomical number. Scientists calculated, for ex-

A muon neutrino like those thought to stream from the Sun creates a horizontal orange line on a computer display tracing its passage through the FermiLab particle detector outside Chicago, Illinois. The event was triggered when a charge-less neutrino entered the detector from the left, striking a target nucleus and unleashing a cone-shaped spray of charged particles. Transformed by the collision, the then-charged muon neutrino traveled to the right, where four green dots reveal its progress through a bank of gallium-filled detectors (below).

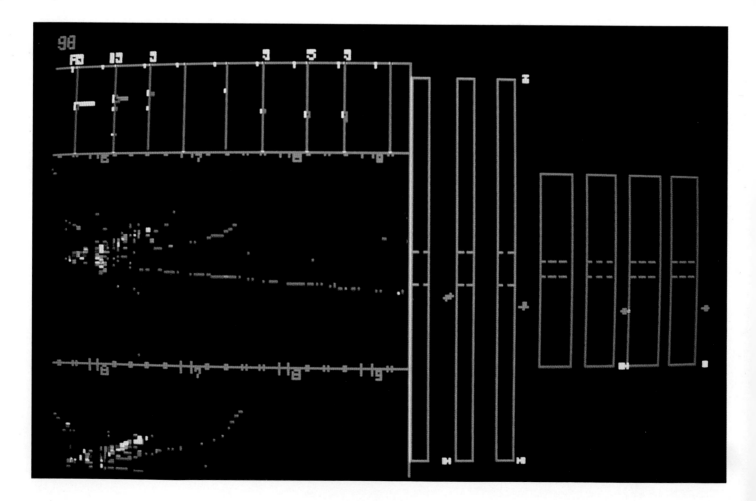

ample, that Earth's share of the Sun's neutrinos—an infinitesimal portion of the total—would amount to 70 billion neutrinos per second for each square centimeter of the planet exposed to the Sun.

Called the neutrino flux, this theoretical number cried out for confirmation. But catching solar neutrinos would be no small feat. An extraordinarily sensitive measuring device would be needed, considering how infrequent antineutrino captures had been in Reines and Cowan's Savannah River experiment, where the antineutrino flux was estimated to be some thirty times the neutrino flux anticipated from the Sun.

TO TRAP A NEUTRINO

Raymond Davis, a physicist then at the Brookhaven National Laboratory on Long Island in New York, and John Bahcall of the Institute for Advanced Study in Princeton, New Jersey, began to investigate the form that a solar neutrino detector might take. The result, in 1967, was the huge tank deep in the bowels of the Homestake mine.

By then, two types of neutrinos were known to exist. One variety, a class of high-energy particles, was called the electron neutrino. Produced by the decay of a neutron into a proton and an electron, this neutrino was the one that Pauli had predicted and that Reines and Cowan had found evidence for. The other, a species of low-energy particle subsequently discovered in the decay of a subatomic particle called a muon, was named the muon neutrino. Scientists had also identified a number of fusion reactions involving elements other than hydrogen, appending no fewer than nine of them to the standard model. Furthermore, theory predicted that each reaction produces neutrinos that vary in energy depending on the elements fused. For example, one of the rarest fusion reactions in the Sun—the joining of a beryllium nucleus with a proton to produce an isotope of boron and a gamma ray—also produces electron neutrinos of exceptionally high energy.

Davis and Bahcall calculated that only these neutrinos would be energetic enough to trigger their neutrino catcher by changing chlorine atoms in perchloroethylene to an isotope of argon, whose atoms could be counted. Taking the relative scarcity of such neutrinos into account, Bahcall judged that the detection of a single such particle per day would correspond to the total solar neutrino flux at Earth's surface.

Perchloroethylene was piped into the reservoir in the spring of 1967. Technicians tested the argon-extraction system, both to make certain that it worked as planned and to cleanse the perc of any argon present before the experiment began. The researchers tended their apparatus much as a hunter would tend a trap—by checking it occasionally to see what it had caught.

At intervals of several weeks, the experimenters bubbled helium through the perc to extract argon atoms for enumeration. Much to the surprise and consternation of the researchers, the tank of perc deep in the Homestake mine detected just one solar neutrino every three days. If that count was accurate, said Bahcall, "there is something wrong either with the

Sun or with the neutrinos—or with what we think we know about them."

He and a quartet of associates carefully analyzed the experiment to confirm that it was based on sound physics and that no one had forgotten to multiply or divide by three, or that a decimal point had not been mispositioned. No such errors could be found, either by Bahcall or by other physicists who investigated. The conclusion: The neutrino deficit was real. For the first time since the discovery of Earth's geologic age had torpedoed the Lane-Emden theory, scientists were confronted with experimental data indicating that their picture of the Sun was flawed.

A MATTER OF ENERGY

As word of Davis and Bahcall's findings circulated within the scientific community, hypotheses to explain the missing neutrinos sprouted like weeds. In 1969, for example, physicists at the Soviet Academy of Sciences saw an answer in particle metamorphosis. Under the direction of Bruno Pontecorvo—an Italian physicist transplanted to the Soviet Union—the scientists expanded on a theory proposed in 1963 by a team of researchers in Japan: Electron neutrinos could become muon neutrinos, and vice versa. Pontecorvo and his collaborators suggested that en route from the Sun to Earth, many of the Sun's high-energy electron neutrinos change into muon neutrinos. Because of their lower energy, the muon neutrinos would be unable to convert a chlorine atom into one of argon and thus would not be counted in the Homestake experiment.

Research into the matter culminated in 1986, when the redoubtable Hans Bethe performed calculations indicating that such oscillations were possible and that the density of material in the center of the Sun might be high enough to induce more than two-thirds of the electron neutrinos produced there to become muon neutrinos during the half-second or so that they tarry in the core. Should this hypothesis prove correct, it would not only explain the missing neutrinos but could also have profound implications for theories about the destiny of the universe.

Cosmologists have long been puzzled by calculations showing that the speeds at which galaxies flee each other are lower than the combined masses of their stars would suggest. The amounts of material visible in the galaxies are simply insufficient for their mutual gravitational attraction to prevent them from flying apart at a much greater rate. To supply this missing mass, cosmologists have proposed that galaxies must contain some kind of "dark matter"—mass in an invisible and as yet unobserved form—that would account for their unexpectedly sedate pace.

Neutrinos might provide the answer. According to nuclear theory, a particle without mass, though it may absorb or emit energy, cannot change form. Thus, for an electron neutrino to become a muon neutrino, these particles must have at least a minuscule amount of mass. The density of neutrinos in the universe is estimated to be between one billion and ten billion per cubic meter. Taken all together, they might be weighty enough to account for the speeds of galaxies. And if the total mass of neutrinos is great enough, its influence could put to rest the idea that the universe will expand forever. Instead, the cosmos will one day begin to contract. After billions of years, its temperature will equal, then exceed, that of the stars. All matter will be vaporized, then reduced to a plasma of dissociated subatomic particles. Eventually, the universe will become a point of pure energy, perhaps to be reborn in another Big Bang.

A first step in testing the theory of neutrino transmutation is to count low-energy muon neutrinos from the Sun's proton-proton reaction to determine whether they, like the high-energy electron neutrinos that Davis and Bahcall counted, are only one-third as plentiful as predicted by the standard model of the Sun. In an effort to capture them, Soviet scientists completed a detector in 1988 at a specially constructed scientific community in the Caucasus dubbed Neutrino Village. The instrument contains sixty tons of gallium, a rare and expensive metal whose atoms can be transmuted into germanium through collisions with low-energy neutrinos.

Whatever the outcome, the Soviet experiment promises to raise as many questions as it answers. If the results show no deficit for low-energy neutrinos, then scientists must still uncover experimental evidence to support the idea that electron neutrinos can change into the muon variety en route from the Sun. Lacking such evidence, astrophysicists may be at a loss to explain convincingly the results of the Homestake experiment, still in operation. Should the shortfall extend to muon neutrinos, the scarcity could mean that the Sun actually produces fewer neutrinos than once thought, compelling other researchers to track down the flaw in the standard model's vision of the Sun's core.

TAKING THE SOLAR PULSE

In the radiative and convection zones that surround the core, there are mysteries just as baffling as those encountered within the Sun's engine. For probing these regions, astrophysicists over two decades have developed a tactic called helioseismology, a technique similar to the seismology that allows geophysicists to construct detailed models of Earth's interior from vibrations caused by earthquakes and other subterranean events. However, instead of sensitive detectors called seismographs to record low-frequency vibrations—subsonic pressure waves—in Earth's crust, helioseismologists use heliospectrographs to take the pulse of disturbances inside the Sun, almost to the depth of the core. With this instrument, an astronomer photographs the Sun in a narrow range of its spectrum. Areas of strong radiation at the chosen wavelength show up brightly in the resulting image, whereas regions of less intensity appear darker.

While investigating the solar surface with such an instrument at the Mount Wilson Observatory in 1960, Robert Leighton of the California Institute of Technology and two colleagues, Robert Noyes and George Simon, noticed that areas of the solar surface that appeared bright in one heliospectrogram

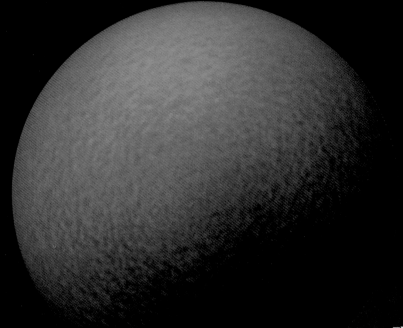

HIDDEN MESSAGES IN SUNQUAKES

To improve their understanding of the Sun's interior, solar scientists have combined methods related in principle to the earth science of seismology with the standard astronomical technique known as Doppler shift measurement. Just as geophysicists are able to glean a great deal of information about the structure and composition of Earth's interior from earthquake vibration patterns that are received at the surface, astronomers can gauge the nature of the Sun's

The wave trail. This cutaway illustration shows how a pressure wave of a particular frequency *(dark bands)* resonates inside the Sun. Generated by turbulence in the convection zone, the wave first travels outward until the sudden change in density at the surface causes it to reflect back toward the interior. The angle at which the wave reflects determines how far it will penetrate: The shallower the angle, the shallower the penetration. Because the speed of sound increases with temperature—which in the Sun increases with depth—the wave's inner edge travels faster than its outer edge, causing it to bend, or refract, toward the surface. The wave will continue this pattern of reflection and refraction all the way around the Sun.

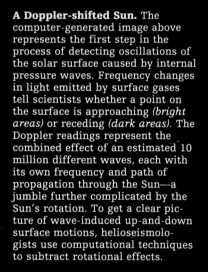

A Doppler-shifted Sun. The computer-generated image above represents the first step in the process of detecting oscillations of the solar surface caused by internal pressure waves. Frequency changes in light emitted by surface gases tell scientists whether a point on the surface is approaching *(bright areas)* or receding *(dark areas)*. The Doppler readings represent the combined effect of an estimated 10 million different waves, each with its own frequency and path of propagation through the Sun—a jumble further complicated by the Sun's rotation. To get a clear picture of wave-induced up-and-down surface motions, helioseismologists use computational techniques to subtract rotational effects.

innards by studying oscillations on the solar disk.

Such disturbances, once thought to be purely localized effects within the photosphere, are actually generated by pressure waves radiating through the interior. Produced by gases bubbling up along massive currents in the convection zone, the very low frequency waves travel outward to the surface, where they cause photospheric gases to rise and fall in a ponderous rhythm, with as much as eleven minutes between wave crests. Upon encountering the sudden drop in density at the photosphere, the waves bounce back toward the interior; reverberating between the surface and subsurface layers, they make the entire Sun vibrate like a bell.

The surface oscillations are observed from Earth by measuring the Doppler shift, or frequency change, of radiation emitted by a particular surface gas. In this way, helioseismologists can determine whether a given point on the surface is moving toward or away from Earth. Computer processing helps to isolate individual waves and their characteristics, which are influenced in recognizable ways by such factors as the temperature, density, chemical composition, and rotation rate of the regions they passed through. By correlating data from a wide assortment of waves—some plunging deep into the interior, others following shallower paths—scientists are building a detailed, layered image of the Sun's insides.

Patterns of oscillation. A computer simulation based on mathematical modeling of just one of the Sun's millions of pressure waves reveals the wave's effects on solar material not only at the surface but throughout the interior. Red indicates wave-induced movement toward the center of the Sun, while blue represents outward motion. Studying many such patterns helps helioseismologists determine variations in density and composition in the Sun's gaseous layers.

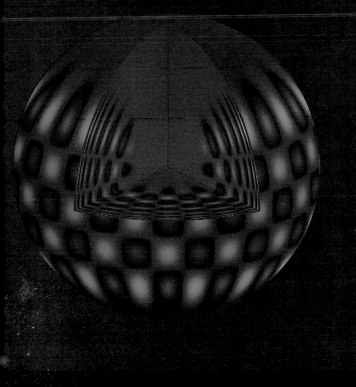

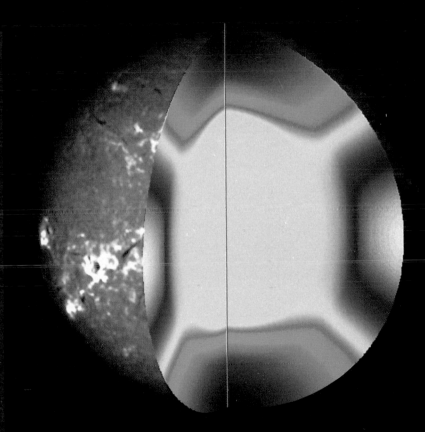

Mapping subsurface features. The colors in this computer image represent differences in rotational speed for various regions of the Sun, ranging from white for the highest speeds to dark blue for the lowest. Helioseismologists determine these rates by analyzing frequency changes in waves traveling through different regions: Wave frequencies either increase or decrease (depending on whether the waves travel with or against the direction of rotation) in proportion to how fast the material they pass through is moving. The resulting profile indicates that interior speeds vary with latitude more than with depth.

became dark in another taken a short while later. The only explanation was that such areas must be bobbing up and down. On the way up, the frequency of light matched the wavelength for which the instrument was adjusted. On the way down, as an area receded from the instrument, the Doppler effect increased the wavelength slightly so that it could no longer be detected. The first satisfactory explanation for the oscillation came in the early 1970s. Roger Ulrich of UCLA—and others working independently—showed that the interior of the Sun could behave much like an organ pipe, which traps and amplifies sound waves by reflecting them repeatedly.

Pressure waves reverberating through the Sun are made up of several million components having wavelengths as short as 1,000 miles and as long as the 2.8-million-mile circumference of the Sun. They travel upward through the convection zone and are reflected back down from the solar surface. Before reaching the core, they are reflected upward again by the high-temperature plasma, which serves as a kind of mirror.

The only way to determine the frequency of a wave is to count the crests of its components, which appear as bright spots, ridges, or rings in a helioseismogram. However, many of these waves differ only very slightly from each other in length—and therefore in the rate at which their crests appear at the surface. Counting enough crests to be certain whether two pressure waves have identical frequencies often requires more time than there are hours of daylight.

One solution, adopted by French astronomers Eric Fossat and Gerard Grec in 1979, was to travel to the South Pole during the austral summer, when

Six of these eleven solar observation sites will form the Global Oscillation Network Group (GONG), whose goal is to map the Sun's interior by keeping its surface oscillations under scrutiny twenty-four hours a day. Although the minimum number of sites needed for continuous observation from the rotating Earth is three, the likelihood of bad weather and equipment malfunctions persuaded GONG astronomers to establish six separate posts at 60-degree intervals around the world.

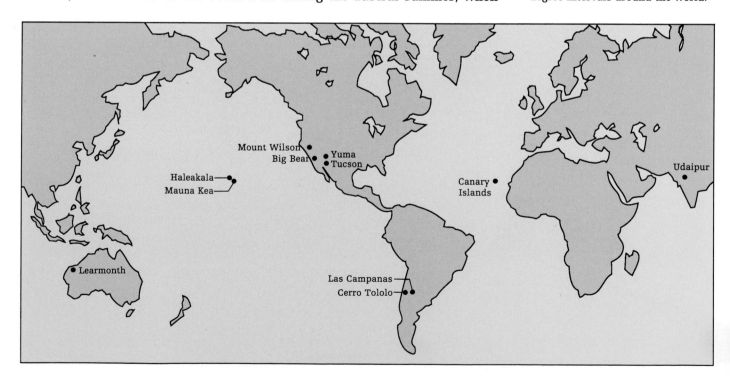

extended daylight allowed them to make continuous observations for almost five days. Another approach is to combine the readings taken at multiple observatories, such as those of the Global Oscillation Network Group (GONG). Astronomers of the National Solar Observatory in Tucson, Arizona, led the way in founding the organization, which has established six solar telescopes around the world. Beginning in 1992, GONG will make two helioseismograms of the Sun each minute, around the clock, for three years. Instead of a heliospectrograph, each of these observatories will use an instrument called an interferometer, widely adopted by helioseismologists for its superior accuracy in gauging the Doppler effect.

Analysis of hard-won seismic data from the Sun has already helped to refine the standard model. For example, Ulrich and others have used such information to revise the depth estimated for the convection zone from 20 percent of the Sun's radius to 30 percent and to set the proportion of helium in the core at 29 percent, slightly higher than that element's 26-percent concentration in the surface layers. Scientists are interested in the amount of helium present in the Sun not only for the effect on the standard model (less hydrogen in the core lowers the temperature there), but also because the quantity of the gas remaining in the Sun is indicative of how much helium the universe had when it was new.

Frank Hill, a solar physicist at the National Solar Observatory, hopes to detect evidence of giant convection cells that not only well up from tens of thousands of miles below the Sun's surface but also appear to move sideways in a manner that might account for observed variations in the strength of the Sun's magnetic field. Suspected of traveling at 100 meters per second, these vast surges of charged gases containing many free electrons would constitute strong electric currents capable of generating powerful magnetism. This in turn could help explain the apparently cyclical character of sunspot activity, for example, or the schedule of solar flares. Should Hill's research settle the question of how gases swirl about inside the Sun, the riddle of its magnetic field may well be solved—if not entirely, then perhaps partly. Alternatively, the results of his experiments may turn out to be as inconclusive and perplexing as the data that have emerged from the Homestake neutrino experiment.

Whatever the answers to puzzles about electromagnetism in the outer regions or about fusion at the core, the Sun's interior will remain a realm of marvels—a hydrogen-fueled furnace large enough to engulf one million Earths, with upwellings of plasma fifty times the diameter of Mars floating toward the surface, stentorian reverberations echoing across distances of hundreds of thousands of miles, and a mountain of matter turned into energy every second. A star—even an ordinary star—is a wondrous thing indeed.

EYES ON EARTH'S HOME STAR

Conventional star though it is, the Sun presents a unique set of observational opportunities and problems. The chief advantage for solar astronomers is the Sun's proximity, which reveals its surface and atmosphere in a degree of detail not possible with more distant stars. Yet the blessing can also be a curse, for solar radiation is much more intense than starlight; its heat can roil the calm atmosphere essential to a clear view from Earth. To balance these benefits and drawbacks, solar astronomers use specialized telescopes, exemplified by those installed in the five observatories presented here and on the following pages.

Due to the brightness of their target, most solar telescopes have smaller apertures than their stellar counterparts, yet they are able to produce large—and extraordinarily detailed—images of the Sun by focusing the light beam over a long span, or focal length. The light path is often encased in a vacuum tube to eliminate the turbulence and consequent distortion that would occur if air molecules were heated by the solar rays; in other cases, instruments are buried underground or fitted with elaborate cooling systems to keep temperatures within the light path stable.

For the best seeing, as observing conditions are known, solar facilities, like other observatories, are often sited at high altitudes, where the air tends to be freer of the dust and water vapor that scatter light. Solar astronomers have another priority as well: long hours of daylight. In pursuit of sites that offer the most consecutive hours of observing time, they travel to the ends of the Earth.

Baillaud Dome *(right)*, Pic du Midi, France

Solar-oscillation telescope, Amundsen-Scott Station, Antarctica

Big Bear Solar Observatory, Big Bear City, California

Vacuum tower telescope, Sacramento Peak, New Mexico

McMath Solar Telescope, Kitt Peak, Arizona

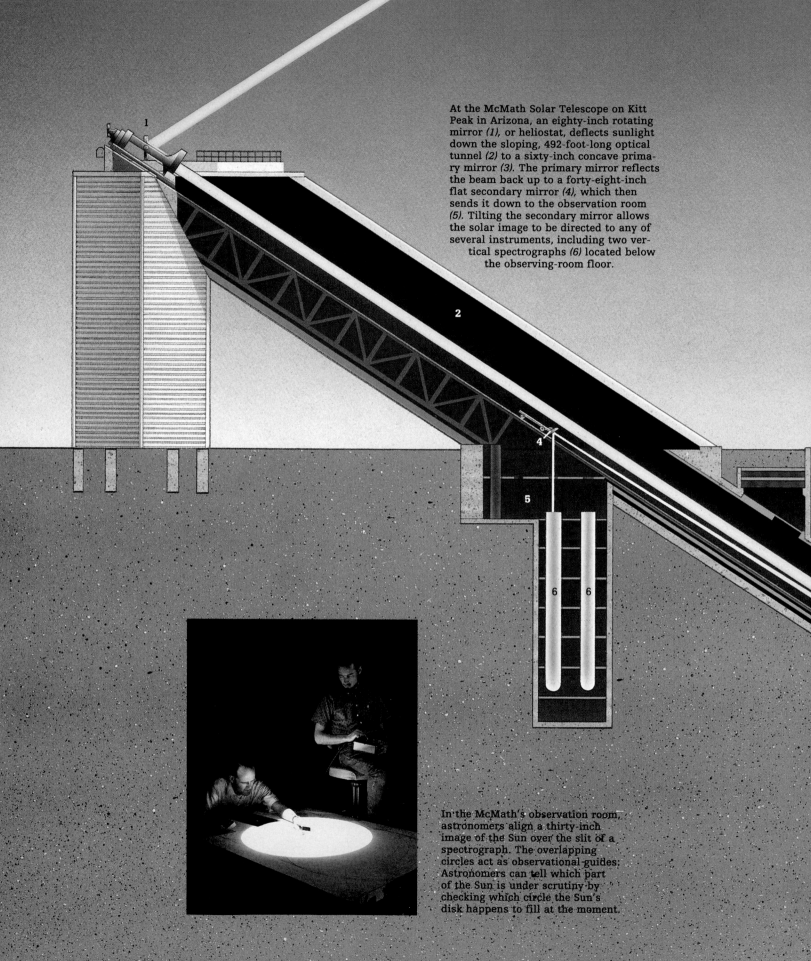

At the McMath Solar Telescope on Kitt Peak in Arizona, an eighty-inch rotating mirror *(1)*, or heliostat, deflects sunlight down the sloping, 492-foot-long optical tunnel *(2)* to a sixty-inch concave primary mirror *(3)*. The primary mirror reflects the beam back up to a forty-eight-inch flat secondary mirror *(4)*, which then sends it down to the observation room *(5)*. Tilting the secondary mirror allows the solar image to be directed to any of several instruments, including two vertical spectrographs *(6)* located below the observing-room floor.

In the McMath's observation room, astronomers align a thirty-inch image of the Sun over the slit of a spectrograph. The overlapping circles act as observational guides: Astronomers can tell which part of the Sun is under scrutiny by checking which circle the Sun's disk happens to fill at the moment.

Bringing the Sun Underground

The dramatic architecture of the McMath Solar Telescope in Arizona is a direct response to a number of technical challenges. The 492-foot-long light shaft makes the McMath the largest solar-observing instrument in the world and also produces the largest images. The shaft inclines at an angle of 32 degrees to the horizon, which orients the telescope parallel to Earth's axis of rotation and makes it easier to track the Sun across the sky. Sunlight is funneled into the shaft by a large rotating mirror known as a heliostat. The mirror, perched atop a 100-foot tower to avoid ground-level air turbulence and blurring of the image, turns automatically with the Sun's passage overhead.

To maintain the telescope's components at a uniform temperature, more than half of the light shaft is buried in the summit of 6,875-foot-high Kitt Peak. The shaft is also encased in a copper skin threaded by a system of coolant pipes that act to stabilize the air within the instrument. Unlike solar telescopes that remove image-distorting air by means of a vacuum system, this arrangement permits the passage of infrared radiation—which is useful for distinguishing weak magnetic fields on the Sun's surface. In the vacuum system *(overleaf)*, a large portion of the infrared is absorbed by a glass window that prevents air from entering the telescope.

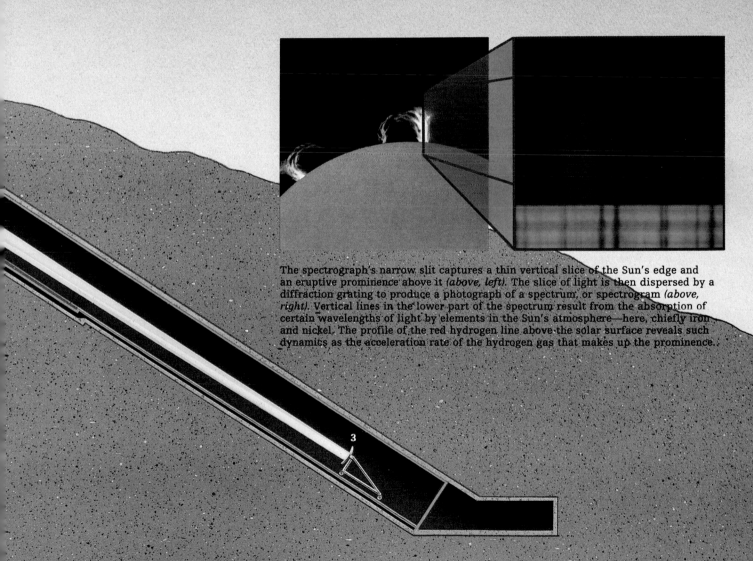

The spectrograph's narrow slit captures a thin vertical slice of the Sun's edge and an eruptive prominence above it *(above, left)*. The slice of light is then dispersed by a diffraction grating to produce a photograph of a spectrum, or spectrogram *(above, right)*. Vertical lines in the lower part of the spectrum result from the absorption of certain wavelengths of light by elements in the Sun's atmosphere—here, chiefly iron and nickel. The profile of the red hydrogen line above the solar surface reveals such dynamics as the acceleration rate of the hydrogen gas that makes up the prominence.

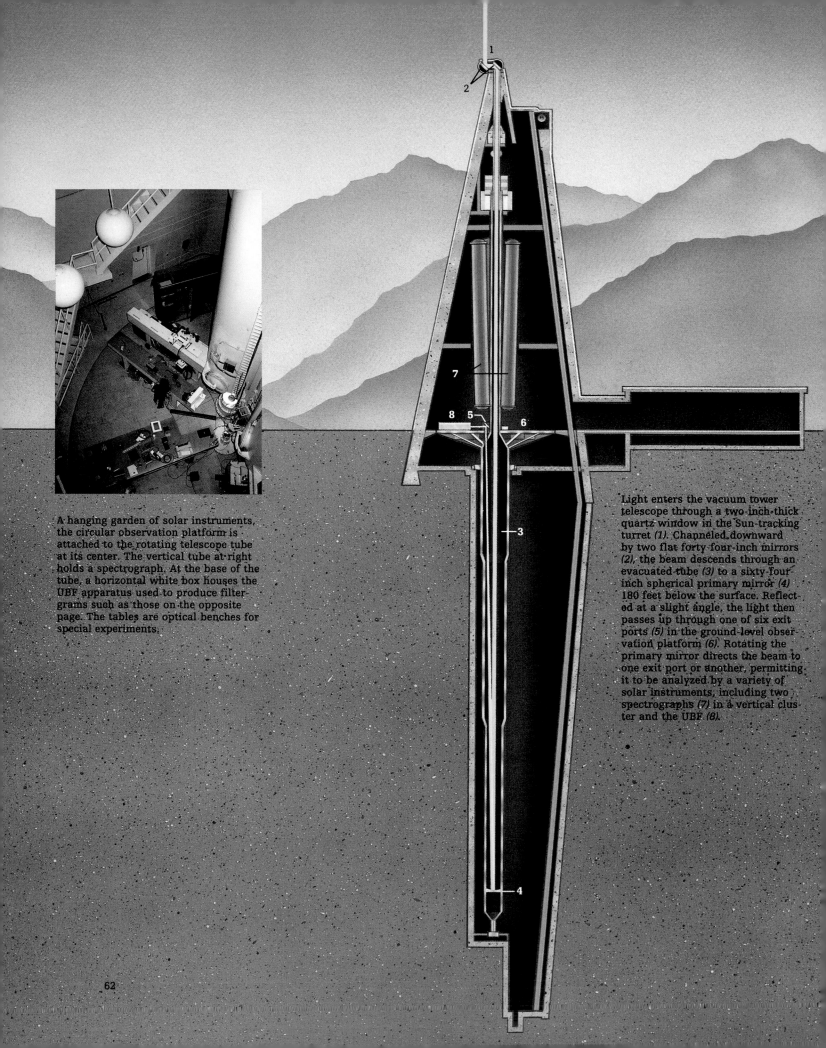

A hanging garden of solar instruments, the circular observation platform is attached to the rotating telescope tube at its center. The vertical tube at right holds a spectrograph. At the base of the tube, a horizontal white box houses the UBF apparatus used to produce filtergrams such as those on the opposite page. The tables are optical benches for special experiments.

Light enters the vacuum tower telescope through a two-inch-thick quartz window in the Sun-tracking turret (1). Channeled downward by two flat forty-four-inch mirrors (2), the beam descends through an evacuated tube (3) to a sixty-four-inch spherical primary mirror (4) 180 feet below the surface. Reflected at a slight angle, the light then passes up through one of six exit ports (5) in the ground-level observation platform (6). Rotating the primary mirror directs the beam to one exit port or another, permitting it to be analyzed by a variety of solar instruments, including two spectrographs (7) in a vertical cluster and the UBF (8).

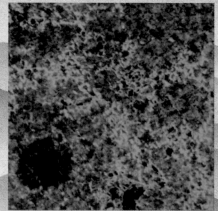

Two filtergrams—the red image *(far left)* produced by light from hydrogen atoms in the chromosphere, the other image *(near left)* produced by magnesium atoms in the photosphere—yield complementary views of two layers of the Sun's atmosphere. Both filtergrams show intense magnetic activity swirling about a large, dark sunspot. By comparing the size, intensity, and motion of such features at various altitudes, astronomers can piece together the three-dimensional structure and dynamics of the solar atmosphere.

A Tower in the Mountains

Located near a precipice on Sacramento Peak in southern New Mexico, the vacuum tower telescope can resolve details on the Sun down to one-fifth of an arc second—an area of the solar surface just 100 miles across. This gives the telescope an effective resolution higher than that of any other solar instrument in the United States. A good deal of the credit for this precision goes to the site. Known to astronomers as Sac Peak, the fir-covered, 9,200-foot-high mountain rises into dry, dust-free air; soft winds ensure a stable air flow and uniform temperatures.

The telescope takes full advantage of this excellent seeing. To bar heated, image-distorting air, the optical path is sealed inside a 321-foot-tall vacuum tube from which almost all the air has been removed. About a third of the tube stands above ground, inside a tower whose three-foot-thick concrete walls are coated with reflective titanium dioxide paint. The rest of the tube extends down a subterranean shaft below the tower.

To keep the image from blurring as the Sun moves across the sky, the telescope and the observation platform containing the analyzing instruments rotate together. The entire 250-ton assembly hangs from a mercury-float bearing near the top of the tower and is turned by an electric motor. So free of friction is the bearing that with the motor off, the suspended bulk can be turned by the push of a finger.

Contributing to the telescope's productivity is a highly sophisticated instrument called the universal birefringent filter (UBF). The UBF admits only a very narrow slice of the solar spectrum at a time and can switch wavelengths rapidly, thus providing nearly simultaneous views of a given region of the Sun at several wavelengths. Since wavelengths of electromagnetic radiation emitted by various elements in the solar atmosphere behave differently at different temperatures, and since solar temperatures vary with distance above the Sun's surface, the UBF's images, called filtergrams, can show an event unfolding at several atmospheric levels. Such three-dimensional representations help elucidate the complex interplay of matter and magnetism on the Sun.

A Mask for the Sun's Face

A special breed of solar telescopes targets the Sun's elusive corona, the halo of white light normally visible only during a total eclipse. The first coronagraph, as such telescopes are known, was installed in 1931 by French astronomer Bernard Lyot at the Pic du Midi Observatory in the Pyrenees. Continued advances in the observatory's instruments since then have kept the Pic du Midi in the vanguard of coronagraphy.

The corona defies casual scrutiny because its faint light is overwhelmed by radiance from the main body of the Sun, which shines a million times brighter. A coronagraph solves the problem by creating an arti-ficial eclipse: Within the telescope tube, a circular plate called an occulting disk blocks the bulk of the Sun, allowing the feeble emissions of the corona to peek out from behind the disk.

As in an actual eclipse, a coronagraph must contend with the sunlight scattered by Earth's atmosphere. Coronagraphs thus are best sited at high altitudes—in this case, a 9,450-foot mountain peak—where the thin, dust-free air keeps the scattering of solar light to a minimum. But even here, high-quality images can be obtained only in the finest weather: Days with low dust and no clouds are dubbed "coronal."

A computer-enhanced image from one of the Pic du Midi's corona-graphs reveals nesting coronal loops, details that help elucidate the dynamics of solar magnetism. The corona is most often observed in an isolated spectral line—here, the iron line—that throws its features into sharp relief.

Bundled against the mountain cold, an astronomer aims the coronagraph located inside the Baillaud Dome of the Pic du Midi Observatory near Tou-louse, France. The dome, which was built in 1907, housed the world's first coronagraph in 1931.

64

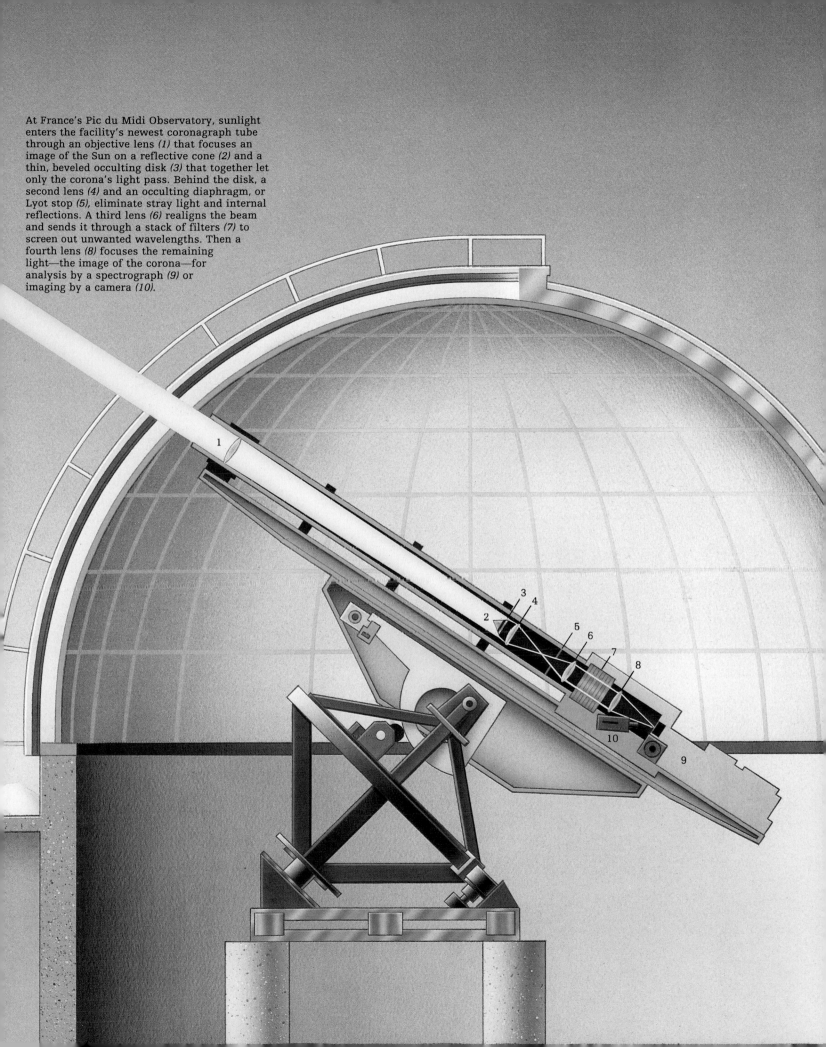

At France's Pic du Midi Observatory, sunlight enters the facility's newest coronagraph tube through an objective lens *(1)* that focuses an image of the Sun on a reflective cone *(2)* and a thin, beveled occulting disk *(3)* that together let only the corona's light pass. Behind the disk, a second lens *(4)* and an occulting diaphragm, or Lyot stop *(5)*, eliminate stray light and internal reflections. A third lens *(6)* realigns the beam and sends it through a stack of filters *(7)* to screen out unwanted wavelengths. Then a fourth lens *(8)* focuses the remaining light—the image of the corona—for analysis by a spectrograph *(9)* or imaging by a camera *(10)*.

BY THE WATERS OF BIG BEAR

Big Bear Solar Observatory, located 6,700 feet high in California's San Bernardino mountain range, offers solar astronomers up to ten hours of good seeing each day. By placing the observatory at the end of a causeway jutting into Big Bear Lake, the facility's designers sought to drastically reduce image-distorting air turbulence: The lake's cool waters absorb much of the Sun's heat rather than reflecting it, as does solid ground, and the lack of obstructions on its flat surface allows wind to flow smoothly. Finally, the site is known for its generally cloudless skies and clean air.

The long daily observing time gives scientists a better chance to spot and track some of the more ephemeral aspects of the Sun's activity, such as surface oscillations *(pages 54-55)* and flares *(below)*, which can change—or even disappear—within minutes or hours. At Big Bear, solar activity studies are made with a trio of telescopes: a twenty-six-inch reflector, in which light is channeled by mirrors, and two refractors, a ten-inch and a six-inch, in which lenses do the job. The smallest telescope monitors the entire disk of the Sun; the larger two—both vacuum systems—examine local regions at increased magnification. Each telescope can be pointed independently, and all three can record their observations simultaneously on an array of instruments.

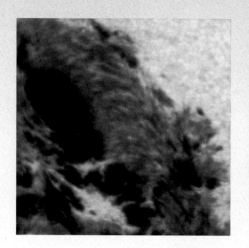 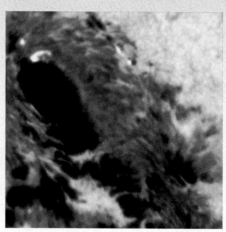 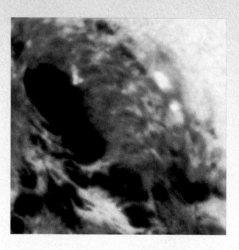

Bright spots denoting solar flares erupt in a five-hour period around a giant sunspot on March 12, 1989. Big Bear's twenty-six-inch telescope captured these images in the yellow spectral line of helium, which reveals solar activity occurring approximately 100 miles above the photosphere.

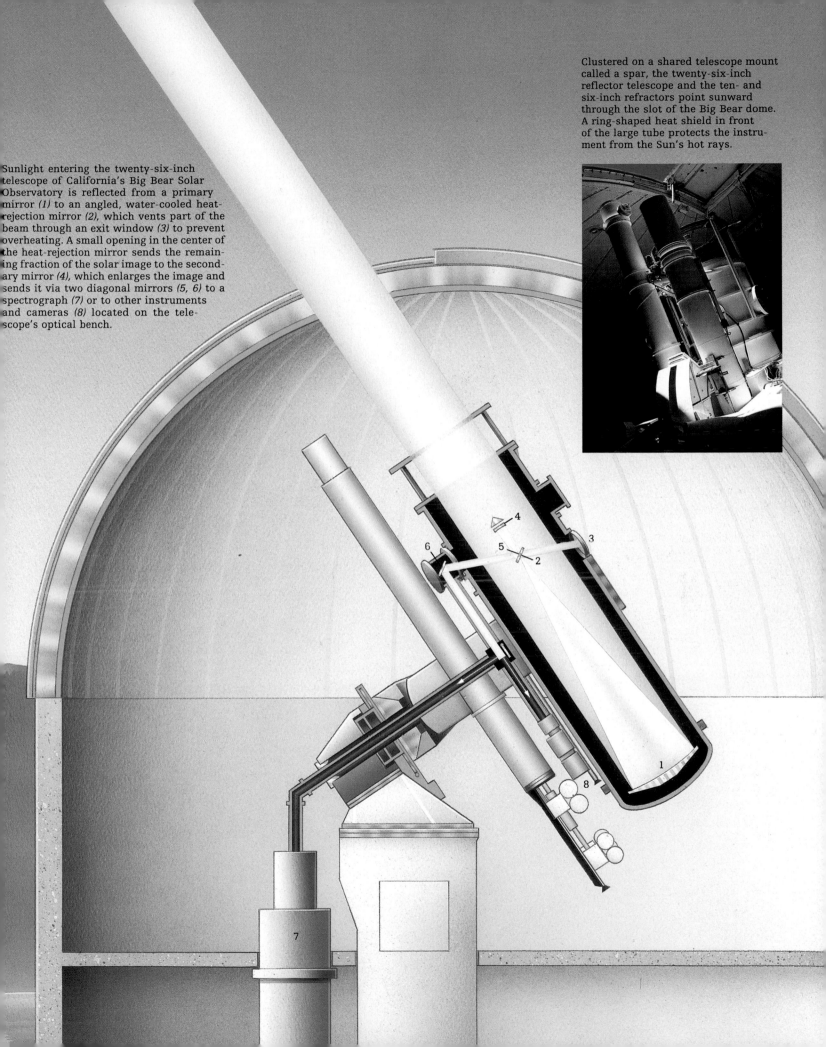

Sunlight entering the twenty-six-inch telescope of California's Big Bear Solar Observatory is reflected from a primary mirror *(1)* to an angled, water-cooled heat-rejection mirror *(2)*, which vents part of the beam through an exit window *(3)* to prevent overheating. A small opening in the center of the heat-rejection mirror sends the remaining fraction of the solar image to the secondary mirror *(4)*, which enlarges the image and sends it via two diagonal mirrors *(5, 6)* to a spectrograph *(7)* or to other instruments and cameras *(8)* located on the telescope's optical bench.

Clustered on a shared telescope mount called a spar, the twenty-six-inch reflector telescope and the ten- and six-inch refractors point sunward through the slot of the Big Bear dome. A ring-shaped heat shield in front of the large tube protects the instrument from the Sun's hot rays.

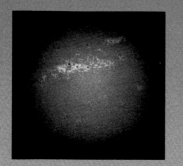

Through filters that admit light emitted by calcium in the photosphere, the Sun's surface appears relatively uniform *(far left)*. But magnified a hundredfold, the contrast between two such images taken 150 seconds apart *(near left)* reveals an oscillating surface: Dark spots are regions where calcium light decreased in intensity, suggesting rising regions; light areas, where calcium increased, denote subsiding regions.

Less than an inch in diameter, the lens *(1)* of the solar-oscillation telescope focuses sunlight through temperature-controlled filters *(2)* onto a high-precision charge-coupled device (CCD) camera *(3)*. The insulated box *(4)* also contains a heated chamber *(5)* for the electronic controls—notably, a signal amplifier-digitizer, which strengthens the images received by the CCD camera and converts them into computer-readable form. Cables beneath the snow bring electrical power to the telescope and carry data to a buried observation room. A wooden fence shields workers from the Antarctic wind.

WATCHERS OF THE MIDNIGHT SUN

A platform located at a high altitude in the cleanest air on Earth would seem to be an ideal site for a solar observatory—especially if the Sun blazed there round the clock for nearly half the year. Just such a setting exists on the icecap that covers Antarctica. Since 1979, solar scientists have made an annual trek to the southernmost continent between October and February to set up their small portable telescopes in a sheltered encampment five miles from the Amundsen-Scott research station, which sits atop the South Pole. Not only does the absence of dust and pollution keep the air crystalline, but at this time of year, the Sun never sets, lingering instead between 12 and 23 degrees above the horizon.

The solar telescope shown opposite, a veteran of three Antarctic summers in the late 1980s, was designed by a consortium of American astronomers to identify patterns in the Sun's surface oscillations, which in turn reveal conditions deep in the star's interior. To accomplish this, the telescope and its attached computer measure changes in the intensity of light emitted by calcium atoms in the Sun's upper photosphere, the layer where the oscillations are most pronounced. Although the observing season is short, the viewing afforded astronomers by the constant daylight allows them to differentiate between oscillation frequencies more accurately than they could anywhere else on the globe.

In an observation room buried beneath four feet of snow, an astronomer keeps track of telescope data on a bank of electronic equipment. The Sun appears as a disk six inches across on the television monitor that rests atop the computer and tape recorder.

69

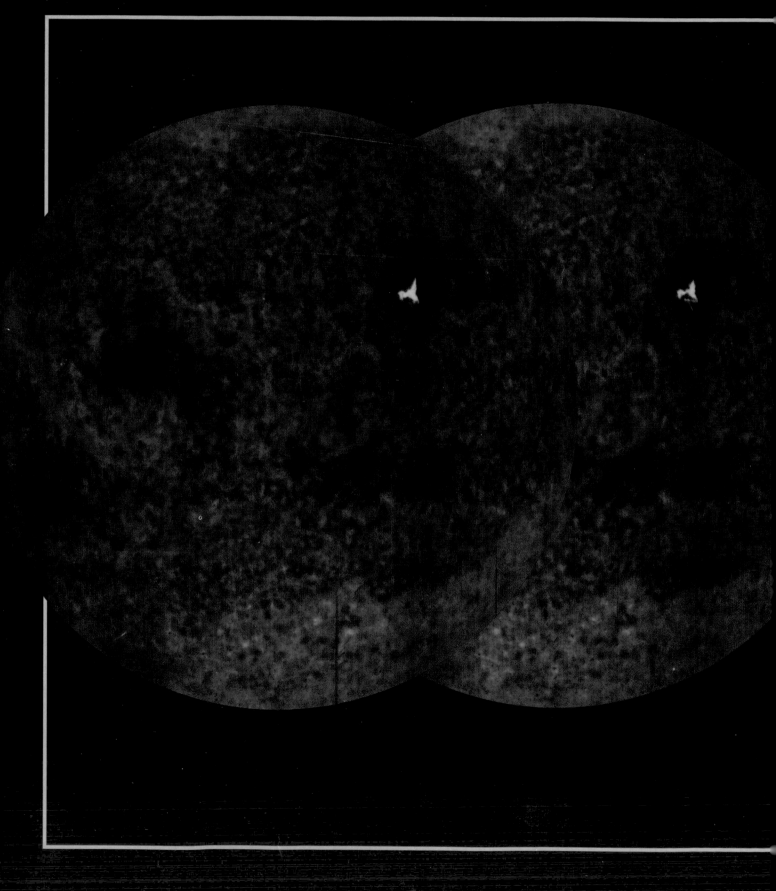

In a series of images taken from Skylab on June 15, 1973, a solar flare brightens and dims like a beacon on the Sun's surface in the space of twenty-two minutes—testament to the almost hair-trigger changeability of Earth's star.

President Ronald Reagan was airborne over the Pacific Ocean en route to the People's Republic of China on April 24, 1984, when an angry outburst from the Sun severed all shortwave communications to and from Air Force One. As the plane's radios fell silent, the Commander in Chief—suspended seven miles high somewhere between Fiji and Midway—found himself unable to send or receive messages. For a period of several hours, the Great Communicator, as Reagan had been dubbed by the press, was cut off from the rest of the world by a star 93 million miles distant. Although this period of presidential isolation ended without mishap, it created some harrowing moments for the Air Force specialists charged with keeping the president in constant touch with military commanders back home.

The x-rays that blacked out communications to the president's plane had originated in a solar flare so intense that measurements of the event went right off the charts. On the specialized scale that solar scientists use to gauge flares, the outbreaks normally range from a low intensity of c-1 to a high of x-10, with the two extremes differing in magnitude by a factor of 1,000. This flare registered a phenomenal x-13, and the radiation storm it generated on Earth did not even begin to subside until the following day.

The Sun has a history of flurries of eruptions alternating with long—sometimes remarkably long—periods of silence. Indeed, the record suggests that the only consistent aspect of this moody star may well be its inconstancy. Flares like the one that isolated the president's plane, for example, are associated with large sunspots, the dark blemishes whose numbers wax and wane in a cycle of approximately eleven years. Yet flares can erupt at any time during the sunspot cycle, and the customary eleven-year span between one solar maximum, or peak of solar activity, and the next can stretch to thirteen years or shrink to a mere eight. And although scientists have recently determined that the Sun dims and brightens slightly, they as yet have no clear idea why, or how, or at what intervals it does so.

TO MEASURE A VARIABLE STAR

Solar scientists first attempted to measure the Sun's luminosity in the early nineteenth century, never dreaming that the star might shine more brightly at some times than others. French observer Claude Pouillet, for example, set out in 1837 to gauge what he called the solar constant—the total amount of radiant energy that the Sun delivers to a square centimeter of Earth in one minute. To do this, Pouillet constructed an ingeniously simple device—a copper box, blackened so as to absorb rather than reflect the Sun's rays, fitted with a thermometer, and filled with water. He called the instrument a pyrheliometer, from the Greek words for "heat," "sun," and "measure." After letting it stand in the shade until the water temperature stabilized, he moved it into sunlight and recorded the subsequent minute-by-minute temperature changes. Modern methods of measuring the solar constant produce results quite similar to those found by Pouillet: His pyrheliometer obtained a figure of 1.76 calories per square centimeter per minute, just nine percent lower than the currently accepted value of 1.96.

Within fifty years, a Yankee inventor with a passion to find out more about the Sun would revise—incorrectly, as it turned out—Pouillet's measurement of the solar constant. Samuel Pierpont Langley, the scion of old and prominent New England families, served as director, or "secretary," of the Smithsonian Institution in Washington, D.C., from 1887 to 1906. During this period, he was a pioneering aviator who even came close to beating the Wright brothers into the air. He had the aerodynamics right, but faulty engines and weak materials sent a number of his prototype machines plunging into the Potomac River.

Before these adventures, Langley worked as a civil engineer, an architect, a college professor, and the director of the Allegheny Observatory at the Western University of Pennsylvania (later, the University of Pittsburgh). It was there, between 1879 and 1881, that he built a solar-radiation-measuring device so sensitive, another astronomer reported, that "the presence of a cow in a pasture could be detected by the heat she radiated, even at a distance of a quarter of a mile." Langley's bolometer (from the Greek word for "ray") consisted of two blackened strips of platinum connected end to end to form a Wheatstone bridge—electrical engineering terminology for a device that measures an unknown electrical resistance by comparing it with a known resistance. When one of the strips was exposed to the Sun, its change in temperature caused the strip's resistance to change; this in turn produced a proportional change in the current flowing through the bridge, which could be measured with such precision that Langley was able to calculate temperature changes of as little as two-millionths of a degree Fahrenheit resulting from a one-second exposure.

Later on, Langley used the bolometer as the detector in a spectrometer whose prism dispersed the Sun's light; in this way, he could measure the intensity of the Sun's radiant energy at all wavelengths then known in the electromagnetic spectrum. To derive a value for the solar constant, he totaled these intensity readings at all spectral wavelengths—a mathematical process known as integrating over the spectrum—and then factored in his estimate of how much solar energy the Earth's atmosphere would absorb.

On July 7, 1881, Langley boarded a train in Pennsylvania with more than two tons of equipment, including his bolometer. Upon arriving in San Francisco fifteen days later, he drafted a retinue of soldiers from a nearby army camp, as well as wagons and mules to cart the gear up the side of Mount Whitney, at 14,495 feet the highest peak in California. "In the still air of this lofty region," he wrote later, "the sunbeams passed unimpeded by the mists of the lower earth."

As it happened, an error in calculating the atmospheric absorption of sunlight above the observing site produced readings for the solar constant that were far too high (they ranged from 2.63 to 3.5 calories per square centimeter per minute), but Langley's sensitive bolometer yielded a monumental discovery nonetheless. While studying the infrared portion of the spectrum, Langley found a region "wholly unknown to science." Now termed the far infrared—with wavelengths ranging from about 30 to 300 microns (a micron is one-millionth of a meter, or four hundred-thousandths of an inch)—this spectral territory has enabled astronomers to ferret out relatively cold objects in the universe, such as dead and dying stars, which may reveal the secrets of stellar evolution.

More progress in the quest to measure the solar constant was made two decades later by a young astronomer named Charles Greeley Abbot. After

Astronomers Samuel Pierpont Langley *(below left)* and Charles Greeley Abbot, shown here with a solar-monitoring device known as a silver-disk pyrheliometer, spent years trying to establish the so-called solar constant—the amount of energy the Sun delivers to a square centimeter of Earth in one minute. Langley discovered the far infrared portion of the solar spectrum, and Abbot compiled important early evidence of the Sun's variable activity.

graduating from MIT in 1895, Abbot had joined the Smithsonian Astrophysical Observatory. Like Langley, Abbot was ahead of his time in a number of ways. His so-called Abbot solar cooker, for example, converted solar energy into heat, allowing him to cook meat and bake bread during the long periods he spent observing the Sun from isolated mountaintops.

In 1903, Abbot learned that the Smithsonian observatory's most recent measurements showed a 10 percent drop in the solar constant. The magnitude of the figure made Abbot doubt its accuracy, so he set out to verify it himself; in so doing, he inaugurated a lifelong quest to derive a precise value for the elusive measurement. "Whether true or false," said Abbot of the reported change, "it was the incentive to the long train of observing which has carried our expeditions to four continents, from sea level to far above the clouds."

The venues for such observations included Mount Brukkaros in present-day Namibia and Mount St. Katherine on the Sinai Peninsula. Small, irregular variations from one day to the next in the solar-constant readings taken at each site led Abbot to conclude that "the Sun is a variable star." But his colleagues resisted the finding, maintaining that the fluctuations might just as easily have been caused by atmospheric distortions or improperly calibrated instruments.

EVIDENCE OF INCONSTANCY

Abbot's belief would eventually be confirmed by space-based solar monitors, yet his contemporaries were justified in viewing measurements of the solar constant as something of a black art. Even though the Sun emits most of its radiation in the visible portion of the electromagnetic spectrum, Earth's atmosphere filters out many of the other wavelengths that contribute to the total flux, or intensity, of the Sun's energy. Molecules of atmospheric ozone, for example, absorb a number of the shorter wavelengths that make up the Sun's ultraviolet light, while carbon dioxide and water vapor screen out many of the longer wavelengths of infrared energy.

Astrophysicists seeking an accurate figure for the solar constant therefore welcomed the wartime birth of rocketry, which enabled them to rise above Earth's atmosphere by proxy. In 1946, for example, solar physicists at the Naval Research Laboratory (NRL) in Washington, D.C., sent a war surplus German V-2 rocket thirty-one miles high to take the first space-based look at the Sun's spectrum. Extraterrestrial measurements of the Sun's energy output, principally in the ultraviolet spectrum, continued in the 1950s with the launch of Aerobee sounding rockets by the NRL, the University of Colorado's Laboratory for Atmospheric and Space Physics, and other laboratories. The monitoring technology developed through these flights—in particular, ultraviolet spectrometers able to operate in a space environment—enabled NASA in 1962 to launch its Orbiting Solar Observatory, or *OSO-1*, the first in a decade-long series of satellites dedicated to observing the solar spectrum. Regrettably, no instruments for measuring the solar constant were included in the OSO series; this left the task to occasional aircraft, rocket, or balloon flights, whose rudimentary instruments and brief observing times were inadequate for detecting changes in the Sun's output.

Not until 1980 did astrophysicists secure definitive evidence of the Sun's variability. On February 14 of that year, NASA placed its Solar Maximum Mission satellite in a planetary orbit 410 miles high. The 4,950-pound craft carried an exquisitely sensitive detector named ACRIM, or active cavity radiometer irradiance monitor, that "revolutionized measurements of the solar constant," according to its creator, Richard Willson of the Jet Propulsion Laboratory in Pasadena, California.

ACRIM measured the solar constant by alternately blocking sunlight from—and admitting sunlight to—a black-walled conical cavity that was kept at a constant temperature by heating coils supplied with a current of electricity. Each time the walls absorbed the Sun's full spectrum of radiation, the wattage required to heat the cavity took a sharp nosedive, and ACRIM faithfully recorded this dip. Every two minutes, the instrument converted its wattage readings into a solar-constant value with an error of just 0.005 percent. ACRIM's measurements established the solar constant at an average of 1,367.5 watts of solar radiation falling on a square meter outside Earth's atmosphere at one astronomical unit—that is, the Earth's mean distance from the Sun. (The superior precision of active-cavity monitoring, as this technique was called, had prompted a shift in the standard of measurement for the solar constant from calories to watts in the late 1960s.)

ACRIM also revealed that the Sun's total energy output varies from day to day. More intriguing still, this luminosity was found to have declined by a total of about 0.1 percent between 1980 and 1986, when it bottomed out and began to rise again. Although climatologists continue to debate the effects of such changes on terrestrial weather, most astronomers no longer question the Sun's variability.

THE SUN CHANGES ITS SPOTS

Above and beyond the inconstancy of its "solar constant," the Sun has shown itself to be a highly variable star in a number of other ways. For many years, astronomers and historians alike maintained that Aristotle's ill-founded belief in a Sun free of imperfections was simply a product of the philosophy of the classical era. In the nineteenth and twentieth centuries, however, evidence began to emerge that the Sun may have gone through extended periods of inactivity in the past—including during Aristotle's lifetime—which could account for an impression of solar serenity.

The German amateur astronomer Heinrich Schwabe had established in 1843 that sunspots come and go in a cycle of 10.4 (later revised to 11) years, yet this reassuring picture of predictability came under attack before the century was out. Studying sunspot distribution in 1887, German astronomer Gustav Spörer noticed that during the years from 1645 to 1715, the number of sunspots reported in the scientific literature from several European countries fell nearly to zero. Although Spörer published two papers describing

In the mid-1970s, John Eddy *(below right)* resurrected work done many decades earlier by Edward Walter Maunder *(right)* and Andrew Ellicott Douglass *(below)*, whose studies suggested that the Sun goes through significant changes in energy output. In the 1890s, when he examined the recorded observations of earlier astronomers, Maunder found long gaps in the eleven-year sunspot cycle. At about the same time, Douglass noted the patterns of tree rings, and Maunder later tentatively correlated those patterns with sunspot changes. The scientific community largely ignored the conclusions of both men until Eddy, shown here with an instrument for measuring the diameter of the Sun, drew together their data with that of other scientists to vindicate their findings.

this anomaly, they were essentially ignored. The one person who did take notice initially suffered the same neglect. In 1890, English astronomer Edward Walter Maunder, the superintendent of the solar department at England's Greenwich Observatory, came across Spörer's findings and decided to make his own search through the same and other records. Maunder saw that Spörer was right: Firsthand accounts of solar observations noted not one spot on the Sun's northern hemisphere between 1672 and 1704. Indeed, fewer sunspots had been sighted during the entire seventy-year period examined by Spörer than were seen during a single average year of the era Maunder was living in.

As it happened, astronomers had not been oblivious to the dearth of sunspots at the time. When a sunspot sighting by astronomer Giovanni Domenico Cassini was written up in a scientific journal in 1671, for instance, the journal's editor pointed out that no other sunspots "have been seen these many years that we know of." Yet when Maunder published his findings in 1894— and in testament to his persevering conviction, again in 1922—the scientific community paid little heed.

Evidence that would ultimately shed light on the case came in the early

1900s from an unexpected source: the study of tree rings. Just before the turn of the century, a twenty-seven-year-old Harvard astronomer named Andrew Douglass moved from Massachusetts to Arizona to become assistant to the director of the Lowell Observatory in Flagstaff. During walks through the evergreen forests that blanketed the surrounding hills, the inquisitive Douglass began to examine the concentric rings in the trunks of felled pines. Not only did the width of the rings vary, he noticed, but the same pattern of narrow and wide rings could be discerned in different trees. This discovery eventually led Douglass to found the modern science of dendrochronology, in which patterns of annual tree rings are used to establish the dates of wood and—especially revealing—of wooden artifacts from various periods in the past. In 1914, after two decades of analyzing cross sections of tree trunks of all ages from all over the world, Douglass announced that tree rings constitute an accurate diary of past weather patterns. In years of drought or dry weather, the trees grow little and their rings, which are faithful indicators of the amount of new woody tissue deposited annually in the outer part of the trunk, are spaced close together; in years of wet weather, by contrast, the trees flourish and their rings are laid down farther apart.

As an astronomer, Douglass was well aware of the eleven-year sunspot cycle, and he began to wonder if it might somehow manifest itself in tree-ring patterns. Reviewing his data, Douglass became convinced that he had found evidence of a connection: an apparent eleven-year cycle that seemed to transcend the rings' year-to-year variability. But he also found something else: There were long periods when the normally varied growth patterns could not be discerned and the tree rings had been deposited in a consistently narrow spacing.

One day in 1922, Douglass received a letter from Maunder himself, reporting the sunspot absence in the period 1645-1715 and asking Douglass to search his tree rings from those years for a uniform pattern—which, Maunder hoped, might prove the validity of his theory about the episode of solar dormancy. Douglass's data set was too small and his analytical techniques too crude to deliver firm statistical proof of Maunder's thesis. Still, the news from England, when combined with his own earlier discoveries, convinced Douglass that solar activity was indeed reflected in tree rings. Like Maunder, however, Douglass failed to sway others, and his attempts to establish a firm correlation between tree rings and sunspot minimums was soon forgotten.

A SKELETON IN THE SOLAR CLOSET

It took an astronomer with a nose for a good detective story to resurrect the deductions of Spörer, Maunder, and Douglass. Solar physicist John Eddy, a senior scientist at the High Altitude Observatory of the National Center for Atmospheric Research in Boulder, Colorado, had always been puzzled by references to the sunspot absence, and in the mid-1970s he set out to prove Maunder wrong. The case of the missing sunspots, felt Eddy, "had hung too long like a skeleton in the closet of solar physics."

Keeping Track of Sunspots

Astronomers have been keeping track of the readily visible spots on the face of the Sun for some 2,500 years. In time, they noticed that the solar blemishes appear and disappear on a roughly eleven-year cycle. (Some researchers now think that the cycles may alternate between ten and twelve years.) Although scientists have yet to determine the precise mechanism that drives the sunspot cycle, they have uncovered several pieces of the puzzle.

The spots appear to be ruled by three aspects of solar physics: First, the Sun is made up of gaseous layers. In addition, it rotates differentially—that is, faster at the equator than at the poles. And finally, it possesses a magnetic field. According to one theory, at the onset of the sunspot cycle, or solar minimum, magnetic field lines run in an orderly pattern from pole to pole through the layers of solar gas. However, the faster-moving equatorial gases soon drag the lines out of their north-south orientation. As hot gases rise from the interior and cooler gases descend from the surface, they create convection currents that pull

the lines further askew. Eventually, the field lines are coiled into tight kinks that break through the bright layer known as the photosphere. Because magnetic fields restrict the motion of charged particles, thereby inhibiting the convection of thermal energy, the places where the lines emerge are cooler than their surroundings and appear as dark spots against the glowing surface.

By monitoring solar cycles, astronomers hope to gain insight into the Sun's internal processes as well as find clues linking solar activity to terrestrial shifts in climate. For example, studies of astronomical and climatic records have revealed an extremely low level of sunspot activity from 1645 to 1715; known as the Maunder minimum, it coincided with a period of cold weather so intense it was dubbed the Little Ice Age. Similar research, combined with techniques such as measuring the carbon-14 content of tree rings, has led some scientists to suggest that Earth has experienced six such minimums in the past 5,000 years—and presumably could be in line for more in the future.

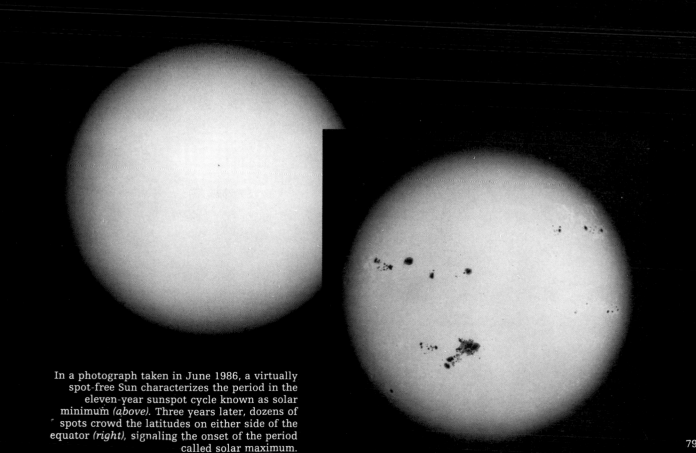

In a photograph taken in June 1986, a virtually spot-free Sun characterizes the period in the eleven-year sunspot cycle known as solar minimum *(above)*. Three years later, dozens of spots crowd the latitudes on either side of the equator *(right)*, signaling the onset of the period called solar maximum.

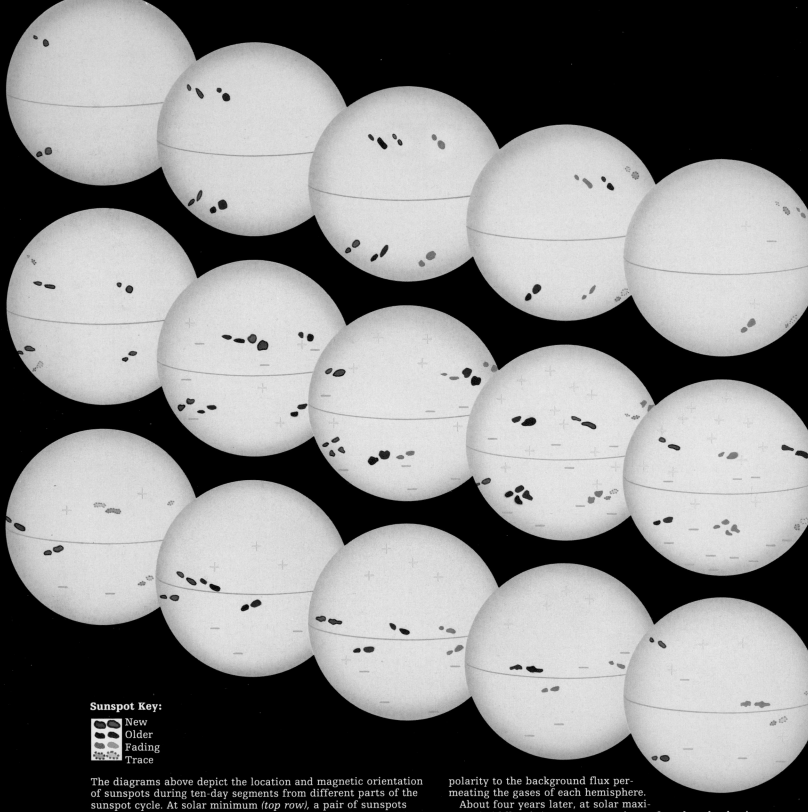

Sunspot Key:

New
Older
Fading
Trace

The diagrams above depict the location and magnetic orientation of sunspots during ten-day segments from different parts of the sunspot cycle. At solar minimum *(top row)*, a pair of sunspots appears around 35 degrees latitude in each hemisphere. In this example, the preceding spot in the northern hemisphere is negative *(red)*, indicating that the north pole has a negative orientation, and the preceding spot in the southern pair is positive *(blue)*; the following spot in each case has the opposite polarity of its mate. As the Sun rotates, new spots continue to emerge at the same latitude and old spots fade, adding remnants of their

polarity to the background flux permeating the gases of each hemisphere.

About four years later, at solar maximum *(middle row)*, spots pepper the surface, but the dominant region of new spots is within 20 degrees of the equator. Areas of positive flux are migrating toward the negative north pole, points of negative flux toward the positive south pole. At cycle's end *(bottom row)*, a scattering of spots persists about 5 degrees from the equator. Then new bipolar pairs appear near 35 degrees, their reversed polarity heralding the start of another cycle.

THE SUN'S MIGRATING MAGNETISM

In the early part of the sunspot cycle, pairs of sunspots—each of which may fade in only a few hours or last for several months—emerge in small groups approximately 30 degrees from the equator in both hemispheres. As the cycle progresses, the spots fade away and new ones appear closer and closer to the equator.

The spots in each pair have opposing magnetic polarities (making the pairs bipolar). The so-called preceding, or leader, spot in each pair, which appears slightly closer to the equator, shares the magnetic orientation of that hemisphere's pole; the following spot has the opposite orientation. About a year after sunspot maximum, the poles reverse their orientation, and the Sun's entire magnetic field begins to reconfigure. Thus, although sunspots wax and wane in the course of eleven years, the Sun's magnetic polarity takes twice that time, or approximately twenty-two years, to cycle through its change.

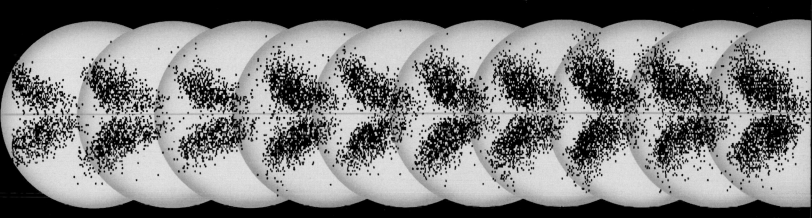

The consistency of sunspot formation is illustrated in the so-called butterfly diagrams shown above. Devised by Edward Maunder in 1904, the diagrams plot the distribution over eleven years of individual spots on the face of the Sun during the ten cycles between 1880 and 1990. In general, spots form first at higher latitudes, then emerge nearer and nearer the solar equator, but the diagrams reveal an overlap of as much as three years between cycles.

The graph below shows the mean number of sunspots and sunspot groups for each year between 1610 and 1980. Although early records are sketchy, the Maunder minimum, from 1645 to 1715, is generally accepted as a significant low in sunspot activity. The large increase in sunspot numbers that lasted from 1710 to 1790, and subsequent rises and dips, lead some astronomers to suggest an underlying solar cycle of about eighty years.

| 1600 | 1620 | 1640 | 1660 | 1680 | 1700 | 1720 | 1740 | 1760 | 1780 | 1800 | 1820 | 1840 | 1860 | 1880 | 1900 | 1920 | 1940 | 1960 | 1980 |

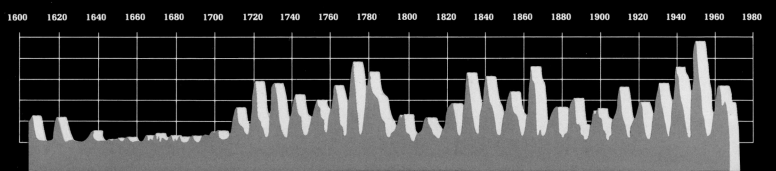

Eddy's fascination with the stars had begun in the early 1950s with a class in celestial navigation at the Naval Academy in Annapolis, Maryland. When he attended graduate school in astrogeophysics at the University of Colorado at Boulder a decade later, however, he was disappointed that the department specialized in the Sun, which he characterized at the time as "a dull star." If he thought so once, his sifting through the literature that had been available to Spörer and Maunder rearranged his thinking soon enough. Eddy reached the same conclusion about the Sun's long period of quiet that his predecessors had, and because he liked the sound of it, he dubbed the sunspot-free era the Maunder minimum.

This time around, the missing sunspots were not ignored, in part because Eddy successfully corroborated his findings with evidence from a number of other fields. Geophysicists, for example, know that numerous and spectacular aurorae can be precipitated on Earth by the outbreak of many sunspots; but in one thirty-seven-year period during the seventy-year Maunder minimum, not a single aurora was reported anywhere around the globe. Further support came from records indicating that no aurorae were seen in London during the sixty-three-year period that ended in 1708, whereas today there would be hundreds in a similar timespan. Even Edmund Halley, the discoverer of the comet that bears his name, scanned the skies for an aurora all his life; born in 1656, he did not glimpse one until after his sixtieth birthday in 1716.

To strengthen his case, Eddy sought confirmation in the astronomical records of the Far East. Sure enough, not one sunspot or aurora had been noted in official records in Japan, Korea, or China between 1639 and 1720, even though astronomers in all three countries had been keeping careful track of such events since the third century AD.

Eddy pulled all this data together in a scientific article published in 1976, pointing out that the lack of sunspot sightings during the period in question could not be attributed to lack of attention. Not only did European astronomers of the seventeenth century have access to telescopes more than powerful enough to see a sunspot, but eminent stargazers such as Johannes Hevelius and Christoph Scheiner had produced sunspot drawings in the 1600s nearly as detailed as the photographic images captured by optical solar telescopes today *(pages 58-69)*.

KEY TO THE CLOSET DOOR

The most convincing evidence for Eddy's theory of prolonged solar silences would come from a renewed—and much more refined—look at tree rings. Since Andrew Douglass's time, scientists had made a number of additional discoveries showing that tree rings reflect not only weather patterns but also the very composition of the atmosphere. For example, when carbon dioxide enters a tree leaf, it becomes part of the new wood deposited in the outer ring of the trunk. Scientists study two forms of this carbon dioxide: an isotope, or variant, of carbon called carbon-12 and the radioactive isotope carbon-14, which is created when high-energy cosmic rays strike nitrogen molecules in

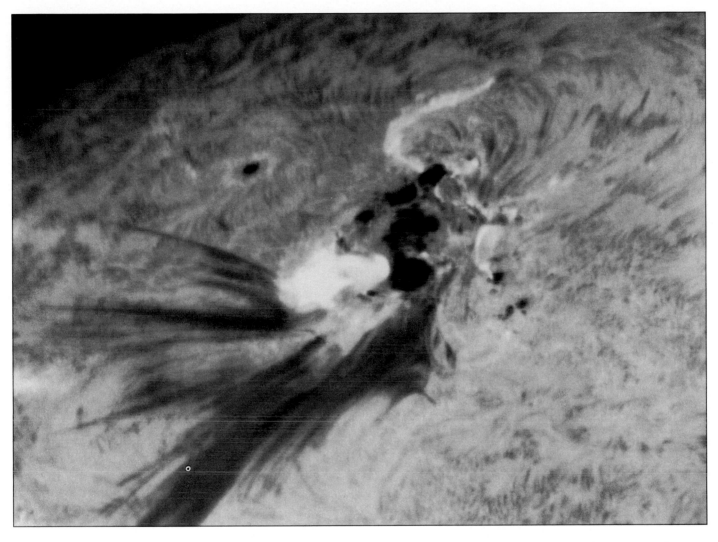

Major solar flares like this eruption from the surface of the Sun in March 1989 release roughly as much energy as a 10-billion-megaton bomb, wreaking havoc on Earth's communications and power systems and generating auroral displays at latitudes where they are otherwise rarely seen. The plume of matter ejected by this flare spanned some 125,000 miles.

the upper atmosphere. Because carbon-14 decays into carbon-12 at a known rate, the ratio of the two isotopes in a tree ring analyzed today reveals how much carbon-14 was present in the atmosphere when the ring was formed.

Eddy was interested in this proportion because when the Sun is active, it emits greater quantities of the charged particles that make up the solar wind; these in turn envelop Earth and deflect many of the cosmic rays that generate large amounts of carbon-14. Periods of prolonged solar inactivity, and the consequent leveling off of the solar wind, should therefore correspond with increased production of carbon-14. Eddy found exactly that: The carbon-14 content of tree rings from the period 1640 to 1720 was markedly higher than in the years either before or after.

Eddy then went on to uncover evidence of other solar minimums—and maximums—in records of sunspot sightings that correlated with tree rings. For example, a prolonged period of sunspot scarcity between 1420 and 1530

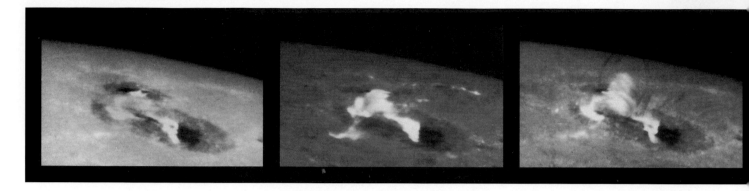

coincided with increased carbon-14 deposits in tree rings from those years. Eddy named this stretch the Spörer minimum in honor of the scientist who first noticed the phenomenon. He also found a period of high sunspot activity, corresponding with low carbon-14 content, in the twelfth and early thirteenth centuries. During this time, which Eddy titled the medieval maximum, glaciers retreated and aurorae lit the night skies over Europe with unaccustomed brilliance and frequency.

Eventually, Eddy turned to the bristlecone pine, the world's longest-lived tree, to trace the record back even further. With the help of measurements of deadwood fragments made by other scientists, he was able to read the history of solar activity in tree rings all the way back to the Bronze Age; again he discovered periodic instances of solar quiescence, each lasting fifty to several hundred years, with the three most recent periods corroborated by astronomical records. From this and other evidence, Eddy concluded that the Sun has spent nearly a third of its existence in a relatively inactive state.

The findings of Spörer, Maunder, and Douglass now seemed confirmed beyond a doubt. "The reality of the Maunder minimum and its implication of basic solar change," Eddy mused before a packed scientific audience in Boston in February 1976, "may be but one more defeat in our long and losing battle of wanting to keep the sun perfect, and if not perfect, constant, and if not constant, regular. Why the sun should be any of these when other stars are not is probably more a question for social than for physical science."

SOLAR FLAREUPS
Of the many questions about the Sun that remain firmly rooted in the domain of physical science, perhaps the most pressing is whether short-term solar behavior can be predicted with any accuracy. If it can, astrophysicists and climatologists will be able to anticipate—and adequately prepare for—the Sun's most dramatic and damaging effects.

The most extensive precautions would likely be taken for a solar flare. As Donald Neidig, one of a team of six Air Force astronomers detailed to the Sacramento Peak Solar Observatory in Sunspot, New Mexico, has noted, "A really big flare can produce enough energy to supply a major city with electricity for 200 million years."

Typically, these monumental flares on the Sun begin as an almost imperceptible bright loop. Within just a few hours, however, the loop suddenly

In the aftermath of a solar flare, material begins to rain down on the Sun's surface, forming so-called loop prominences that follow the lines of magnetic force arcing between two sunspots; the brightest, densest material is at the top of the curves.

explodes, spewing billions of tons of gaseous material—as well as a wide range of electromagnetic radiation and energetic nuclear particles—into space. Eight minutes later, a strong blast of x-rays and ultraviolet rays reaches Earth and radically alters the ionization structure of the planet's upper atmosphere; this in turn plays hob with the way radio waves are reflected from that layer. Although only a fraction of the flare's ejecta ever gets to Earth, it can be enough to disrupt communications and electrical power systems all over the planet. Twenty-four minutes or so later, Earth's neighborhood is bombarded by potentially dangerous high-energy protons traveling at one-fourth the speed of light. Astronauts en route to or from the Moon or Mars, for example, and thus beyond Earth's magnetosphere—the protective sheath of gases, electrical currents, and magnetic fields that surrounds the planet—could suffer lethal radiation poisoning. The last assault is a magnetic shock wave, a fast-moving magnetic disturbance created when matter is expelled from the Sun sometimes at more than 600 miles per second; it washes over Earth one to two days after the flare's eruption.

Rarely has the power of a solar flare been more dramatically displayed than in early March 1989. Over a period of ten days, a series of violent flares unleashed a combined shower of radiation, energized particles, and magnetism that knocked out electricity all across the province of Quebec, rendered normal radio frequencies unusable, and draped the night skies of the Northern Hemisphere with a crimson aurora borealis that could be seen as far south as Key West, Florida.

As the flares' extreme-ultraviolet flux heated and expanded Earth's upper atmosphere, the increased atmospheric drag reduced the orbital energy of hundreds of satellites in low Earth orbit. This knocked the spacecraft into lower and faster orbits, causing ground controllers to temporarily lose contact with them. Meanwhile, many of the 7,000 orbiting objects that are tracked by the U.S. Space Surveillance Network were lost from view.

The U.S. Department of Defense in general, and the Air Force in particular, takes an active interest in studying solar flares because of the threat posed to the services' primary lifeline, radio communications. The blackout that crippled President Reagan's communications in 1984, for example, resulted when high-energy radiation from the flare—primarily ultraviolet and x-ray wavelengths—broke apart individual atoms in Earth's ionosphere, increasing the density of charged particles in the upper reaches of that layer to such a degree that the ionosphere absorbed rather than reflected shortwave radio signals at frequencies between 3 and 30 megahertz. However, the same conditions can sometimes enhance the transmission of very high frequency (VHF) radio signals (between 30 and 300 megahertz), which are normally limited to a range of less than 500 miles. On March 13, 1989, for example, a radio amateur in Rhode Island was able to contact a second operator in England using the VHF band of 50 megahertz.

Perhaps the most extensively studied aftereffects of a solar flare are the spectacular aurorae that appear as sheets or bands of light hung in folds across the sky, where they flicker and pulsate for a few minutes or glow steadily for hours. Although aurorae normally materialize only in high latitudes, within about 20 degrees of Earth's magnetic poles, particularly intense northern lights have been observed as far south as Cuba, Mexico City, and even Singapore, which lies just 1 degree north of the equator. The aurora australis—the Southern Hemisphere's counterpart—has occurred as far north as Samoa, 10 degrees south of the equator.

ELUSIVE QUARRY

Despite a flare's readily apparent effects on Earth, its driving mechanism remains largely hidden from astronomers' view. One imposing obstacle to a clearer understanding of a solar flare is Earth's atmosphere, which allows only the visible wavelengths of a flare's radiation, along with some radio frequencies, to get through. The flare's x-ray, gamma ray, and most of its ultraviolet emissions can be measured only from space.

On May 14, 1973, NASA provided the means for such measurements with its launch of Skylab, the first human-occupied orbiting solar observatory. Until that day, astrophysicists had been forced to rely on a miscellany of rockets, weather balloons, and satellites whose unsophisticated monitors and

In a sequence of photos made during a space shuttle *Challenger* mission in April 1984, astronauts George Nelson and James van Hoften, aided by the rest of the *Challenger* crew, perform history's first orbiting repair job on a satellite, the Solar Maximum Mission *(SMM)*. The rescue began with Nelson's unsuccessful attempt to dock with the *SMM (first two pictures)* but ended triumphantly with the capture, refitting, and redeployment of the satellite, an important tool for monitoring solar changes.

brief forays above the atmosphere provided only tantalizing glimpses of solar flares. Skylab's Apollo Telescope Mount *(page 116)*, in contrast, bristled with twelve tons of observing instruments whose gaze at the Sun would be interrupted only by the portion of each orbital period that the spacecraft spent in Earth's shadow. By witnessing the birth of a flare, solar researchers hoped to find the keys to its workings. The Sun was not a cooperative subject, however. Said Skylab astronaut and scientist Edward G. Gibson, "Every time we realized that a flare was taking place, and started the instruments, we were too late to observe the initial energy release."

The launch of the Solar Maximum Mission *(SMM)* satellite in February 1980, just before the climax of a sunspot cycle, gave scientists a more reliable means for witnessing a flare from beginning to end. Of the seven instruments aboard the satellite, six were dedicated to examining solar flares. However, within nine months of the craft's entering orbit, three fuses had blown in the attitude-control module essential to keeping the *SMM* properly oriented. Although the crippling of the satellite disappointed solar physicists, the *SMM* could not be written off just yet; its designers had anticipated that it might one day need repairs in space.

The astronaut-mechanics who eventually made those repairs reached the *SMM* in a space shuttle, but that mission did not take place right away. At the time of the satellite's impairment, the maiden shuttle launch was still five months away, and for three years after the first shuttle got off the ground in April 1981 it was used as a launch pad for new satellites, not a repair bay for old ones. By the spring of 1984, though, NASA had deployed enough satellites to consider putting shuttle astronauts to work as salvage experts.

On April 8, a five-member crew brought the shuttle *Challenger* into an orbital rendezvous just 100 feet from the stricken satellite. After donning a spacesuit and a nitrogen-powered jetpack known as an MMU (for manned

maneuvering unit), astronaut George Nelson flew untethered to the *SMM*, where he tried to latch an attachment device onto the slowly rotating satellite. Unexpectedly, the effort failed; a cluster of one-eighth-inch-thick fiberglass pins holding the *SMM*'s gold-coated thermal insulation in place kept the device from locking into position.

In frustration, Nelson attempted to grapple the satellite by hand, but that only sent it tumbling out of control. Fortunately, ground controllers at the Goddard Space Flight Center in Greenbelt, Maryland, stabilized the craft long enough so that, two days later, astronaut Terry Hart was able to grasp it with the shuttle's robotic arm. Hart then guided the *SMM* into the shuttle's cargo bay, where Nelson and fellow mission specialist James van Hoften took only forty-five minutes to replace the lifeless attitude-control module. This gave the astronauts ample time to replace another failed component, this one in the *SMM*'s coronagraph-polarimeter. The entire team of astronauts then gathered in the shuttle's crew compartment to celebrate their eventful day with a steak dinner. The next day, having verified that the repairs were successful, the astronauts redeployed the satellite from the shuttle's payload bay.

Unfortunately, the Sun itself brought the *SMM*'s lifetime to a premature end. The increased atmospheric drag that had begun in concert with the flares of March 1989 pulled the satellite so close to Earth that it later fell into the atmosphere and broke apart upon reentry.

Still, the Solar Maximum Mission satellite had a productive life, and like Skylab, it contributed key pieces to the present understanding of the mechanics of flares. The two spacecraft revealed that flares act like colossal atom smashers, accelerating protons or electrons to energies of tens of millions of electron volts. When these particle beams collide with atoms of hydrogen, helium, and other elements in the Sun's chromosphere, they produce high-energy x-rays and gamma rays as well as showers of energized electrons and protons.

Space-based observations have also helped solar physicists piece together the origins of a flare. According to the leading hypothesis, a flare results from a fast, catastrophic rearrangement of the magnetic fields in an active region of the Sun's corona. Somehow, this instability in the magnetic fields accelerates beams of protons or electrons, which follow the corona's arcing magnetic field lines down into the denser reaches of the chromosphere. There, the particle beams are slowed by collisions with protons and free electrons, a process that releases high-energy x-rays and gamma rays and causes superheated plasma to explode into the portion of the corona where the flare began.

Solar scientists lack an instrument sensitive enough to measure the exact strength of the coronal magnetic fields that spawn a solar flare, but the power of the fields can be extrapolated from the strength of even mightier fields measured in the photosphere. In addition, x-ray spectrographic studies of flares have found evidence of iron atoms with twenty-five of their twenty-six electrons stripped away; applying the laws of atomic physics, astronomers have calculated that such a degree of ionization can occur only when tem-

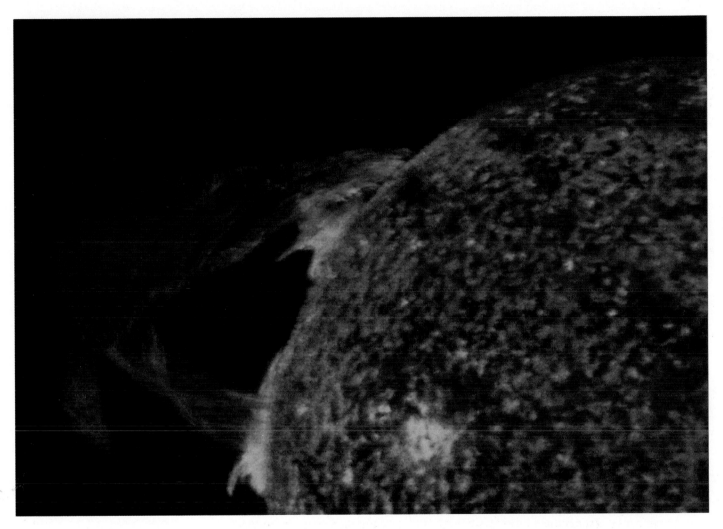

Rising on twisted arcs of magnetic force, a spectacular eruptive prominence photographed from Skylab in 1973 balloons 250,000 miles into space, roughly the distance between Earth and the Moon. Although it was one of the largest such prominences seen in a decade, astronomers noted that its hottest regions reached temperatures of no more than 70,000 degrees Kelvin, many times cooler than the Sun's corona.

peratures approach 50 million degrees Kelvin—three times hotter than the center of the Sun itself. Yet the magnetic fields governing the creation of a flare are strong enough to keep these intensely hot gases from expanding.

COUSINS OF THE FLARE

Although they occur more often and cause less damage on Earth than flares, the related solar outcroppings known as prominences can be every bit as sensational. The smallest solar prominences have widths roughly equal to that of Earth; the largest may approach half the diameter of the Sun itself. These arches of glowing gas float tens of thousands of miles above the solar surface, suspended by the Sun's looping magnetic fields and sculpted into such shapes as hedgerows, funnels, and arcades. Some prominences resemble curtains of vertical ropes—delicate, dancing rivulets that stretch nearly 3,000 miles high. Others look like neat coils that spiral from one spot on the Sun's surface to another 50,000 miles away.

Their behavior varies as widely as their appearance. Some prominences may occupy the same place above the solar surface for months without major change, while others turn slowly in place under the influence of the Sun's differential rotation. Sometimes one half of a prominence will rise while the other falls, and what began as a near-vertical sheet of electrified gases can slowly shred apart into myriad fine threads that flow back to the solar surface. During total eclipses, prominences looming above the rim of the Sun are a brilliant magenta against the inky backdrop of space; those seen against the Sun's main body, or disk, always appear as thin, dark ribbons known as filaments.

But prominences are not always so stately and self-contained. Any prominence that rises more than 30,000 miles above the Sun's surface, for example, is likely to burst within forty-eight hours. Most of these eruptive prominences, as they are called, fling their gases outward, tearing apart the overlying corona and injecting enormous quantities of plasma into space; others, however, simply fall as a kind of rain into the chromosphere. One fascinating aspect of eruptive prominences is that roughly two-thirds of them re-form in the same shape and place several times, suggesting that the explosive releases are a normal part of their life cycle and that special conditions favorable to the formation of a prominence prevail at particular sites.

Many a prominence takes shape in tandem with a solar flare. During a particularly violent flare in August 1972, for example, a giant loop prominence—a sharply defined, symmetrical arch resembling a horseshoe—rose into view over the flare's point of origin on the Sun. Eruptive prominences have also been seen to burst at the same time as a flare, although it is not clear which event prompts the other, or whether both stem from the same motive forces. "An eruptive prominence could be the signature of an instability that's beginning to take place and that ends in a flare," suggests solar physicist Jack Zirker of the Sacramento Peak observatory.

A number of astronomers have proposed that loop prominences reveal the Sun in the process of healing the wounds inflicted by a flare. In this scenario, a loop prominence is a sort of gigantic solar suture, showing exactly where the magnetic field lines pulled apart by a flare are slowly being stitched together again. Once the magnetic reconnection is complete, the prominence fades and disappears.

It would be in keeping with the Sun's imperfectly understood nature if neither of these explanations turned out to be quite right. Pending the next breakthrough in solar observation or theory, then, flares and prominences will continue to pose a beautiful mystery.

GIFTS OF THE SUN

Without the radiance of the Sun, Earth's several trillion tons of living matter—distributed among a million-plus species ranging from single-cell bacteria to world-roving whales—would cease to exist. In a process as complex as it is commonplace, the reaction between sunlight and green plants produces the building blocks of sugars and starches that feed not only the plants themselves, but also the creatures that feed on the plants and the animals further up the food chain that feed on plant eaters. Known as photosynthesis, the process took more than a billion years to evolve, and at first it was a planetwide disaster. One of its chief by-products is oxygen, a lethal gas to Earth's earliest organisms. The blue-green algae that began the photosynthetic experiment thus not only killed off most of their own kind with their noxious exhalations but also sent nearly all other forms of primitive life scurrying for shelter in anaerobic swamps and soils. However, the species that adapted gave rise to almost every living thing today—and photosynthesis supplies the very breath of life.

The Sun offers the promise of even greater gifts to the species called *Homo sapiens:* The natural light that works microscopic magic within a green leaf can also be made to generate electricity to power anything from a pocket calculator to a space station. Earth, orbiting 93 million miles from the fountain of energy at the center of the Solar System, receives only one-half of one-billionth of the Sun's vast output, and more than half of that never reaches the planet's surface. Yet even the tiny trickle that does arrive, if it could be harnessed efficiently, could fulfill the energy needs of the entire globe, with power to spare.

FROM SUNLIGHT TO FOOD, A CHAIN REACTION

By the early 1800s, scientists understood photosynthesis in a general way; they knew, for example, that plants give off oxygen and require light and carbon dioxide to grow. However, it was not until the 1930s and the advent of techniques to isolate radioactive isotopes that biological researchers could begin to discern the step-by-step choreography of the process. As depicted in simplified form here and on the next few pages, photosynthesis involves two sets of reactions. Both take place within plant cell components called chloroplasts, which contain the molecules necessary to carry out photosynthesis, and both are made up of several steps.

The first set of reactions occurs inside grana *(below)*, membranous structures densely packed within chloroplasts. Grana, made up primarily of fat molecules called lipids *(white)*, also contain chlorophyll *(green)* and carotene-like substances *(orange)*, both bound to large protein molecules that are unique to the grana. The chlorophyll absorbs the Sun's energy to begin a chain reaction that produces two chemical compounds, NADPH and ATP *(box, opposite)*. In the course of these reactions, molecules of water taken up by the plant are split, liberating atoms of hydrogen and oxygen. Fueled by NADPH and ATP, the second set of reactions *(pages 94-95)* takes place just outside the grana but still within the chloroplast to give rise to the sugars and starches essential to plant life.

Two Ways to Make Plant Fuel

Inside the grana, two distinct processes—photosystem I and photosystem II—produce the ingredients needed for the second stage of photosynthesis, which is known as the Calvin cycle *(overleaf)*. Both begin when a ray of sunlight, or photon, strikes a chlorophyll molecule, knocking loose an electron and setting in motion a chain of chemical events. In photosystem I, two molecules result: NADPH, formed when an atom of hydrogen joins NADP (nicotinamide adenosine di-

nucleotide phosphate), a substance present in the plant cell; and adenosine triphosphate, or ATP. In photosystem II, a higher-energy photon strikes a chlorophyll molecule, releasing an electron that breaks a molecule of water (H_2O) into its components. The oxygen escapes into the air; one hydrogen atom contributes an electron to replenish the one lost by the chlorophyll in photosystem I, and the other atom aids in the production of ATP.

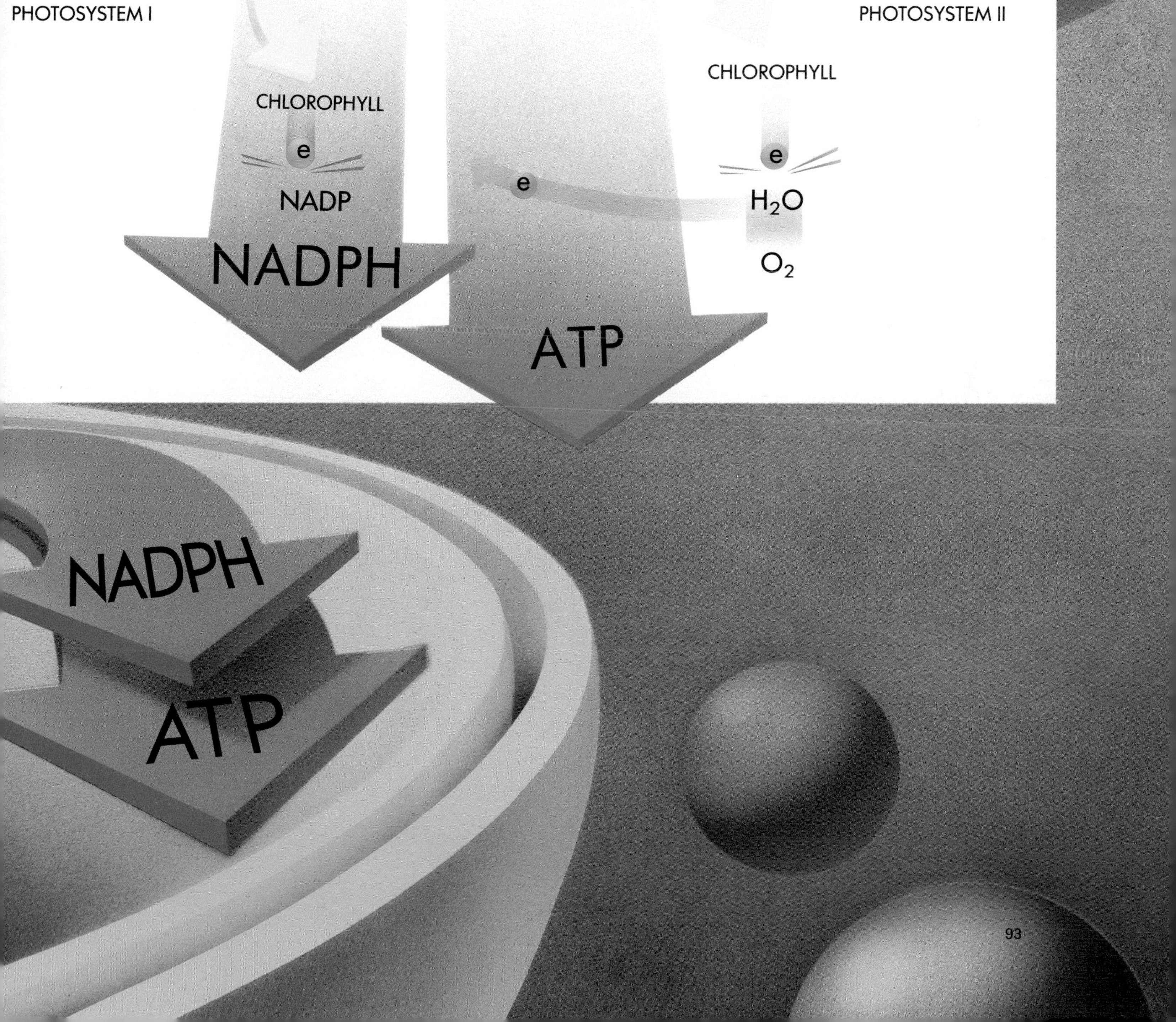

PHOTOSYSTEM I

PHOTOSYSTEM II

CHLOROPHYLL

e

NADP

NADPH

CHLOROPHYLL

e

e

H_2O

O_2

ATP

NADPH

ATP

A Cycle of Transformations

The compounds manufactured by sunlight, chlorophyll, and other biochemicals in photosystems I and II emerge from the grana to take part in a set of reactions—greatly simplified here—often referred to as the Calvin cycle after its principal discoverer, Nobel Prize-winning chemist Melvin Calvin. In this stage of photosynthesis, carbon dioxide, ATP, and NADPH combine in various ways with other substances in plant tissues to produce molecules of the carbohydrates that supply the energy plants need to grow.

The chemical events that make up this reaction cycle were identified by Calvin and a research team at the University of California in a series of experiments begun in the mid-1940s. The team exposed several different samples of green plant cells to an atmosphere rich in radioactive carbon dioxide, then interrupted the process of photosynthesis after varying amounts of time had elapsed by immersing the samples in alcohol: One group of cells was plunged into an alcohol bath after just five seconds, another after a minute, and so on. By studying each sample to determine precisely which molecules had acquired radioactive carbon atoms at which stage, Calvin's team established the step-by-step order of the creation and breakdown of the chemical compounds vital to life.

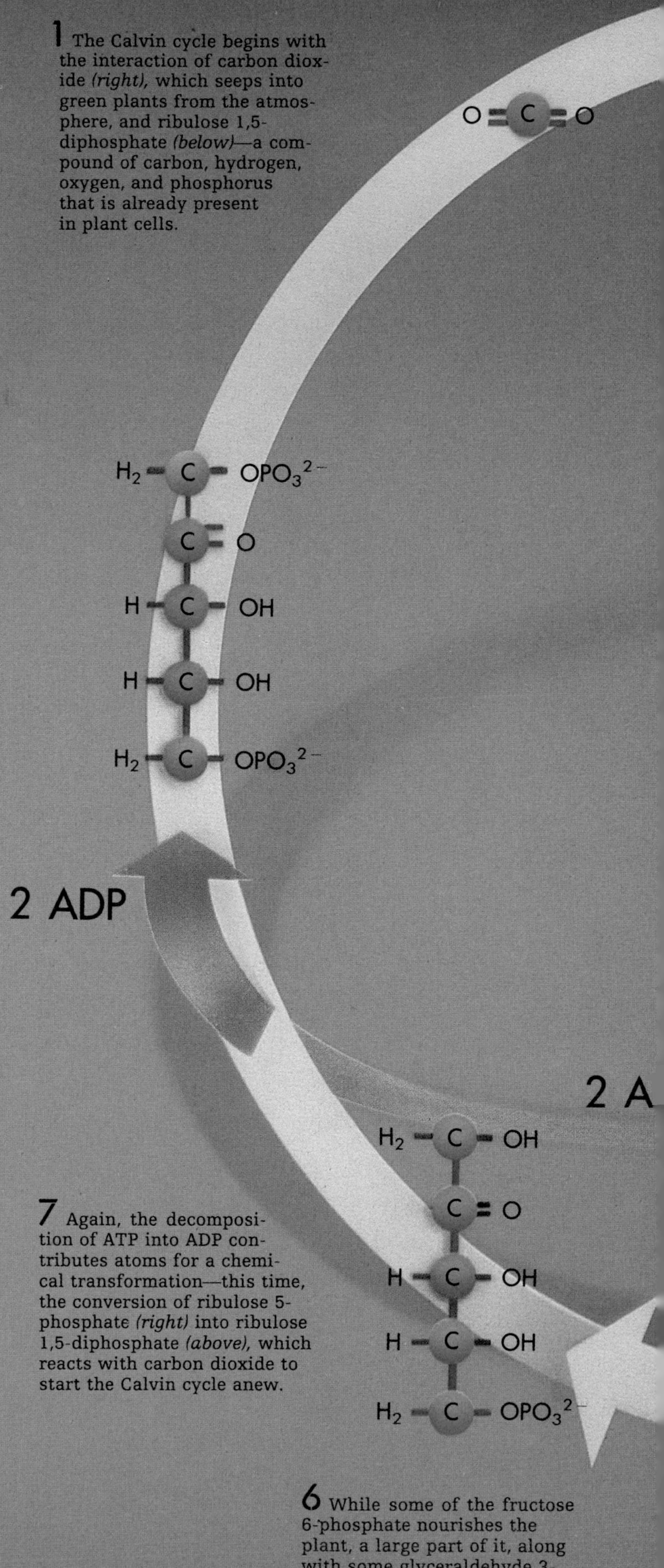

1 The Calvin cycle begins with the interaction of carbon dioxide (right), which seeps into green plants from the atmosphere, and ribulose 1,5-diphosphate (below)—a compound of carbon, hydrogen, oxygen, and phosphorus that is already present in plant cells.

$O = C = O$

$H_2 - C - OPO_3{}^{2-}$
$C = O$
$H - C - OH$
$H - C - OH$
$H_2 - C - OPO_3{}^{2-}$

2 ADP

2 A

$H_2 - C - OH$
$C = O$
$H - C - OH$
$H - C - OH$
$H_2 - C - OPO_3{}^{2-}$

7 Again, the decomposition of ATP into ADP contributes atoms for a chemical transformation—this time, the conversion of ribulose 5-phosphate (right) into ribulose 1,5-diphosphate (above), which reacts with carbon dioxide to start the Calvin cycle anew.

6 While some of the fructose 6-phosphate nourishes the plant, a large part of it, along with some glyceraldehyde 3-phosphate, undergoes a series of reactions with compounds called enzymes to produce ribulose 5-phosphate (above).

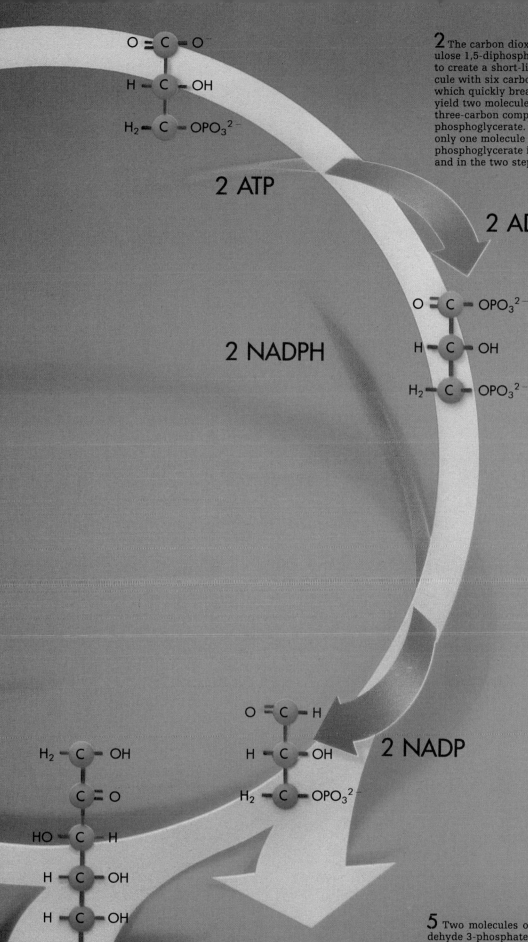

2 The carbon dioxide and ribulose 1,5-diphosphate combine to create a short-lived molecule with six carbon atoms, which quickly breaks down to yield two molecules of the three-carbon compound 3-phosphoglycerate. (For clarity, only one molecule of 3-phosphoglycerate is shown here and in the two steps that follow.)

2 ATP

2 ADP

2 NADPH

3 Next, two molecules of ATP decompose into two molecules of ADP (adenosine diphosphate), giving up phosphate ions that attach to the molecules of 3-phosphoglycerate to create 1,3-diphosphoglycerate.

4 As hydrogen atoms from two NADPH molecules enter the reaction cycle, they convert the 1,3-diphosphoglycerate to molecules of glyceraldehyde 3-phosphate and break down the NADPH to NADP. Some of the glyceraldehyde 3-phosphate leaves the cycle and is transported out of the chloroplast to make complex sugars for exportation to the leaves, roots, and fruits.

2 NADP

5 Two molecules of glyceraldehyde 3-phosphate (again, only one is shown) combine in a series of steps to form one molecule of the carbohydrate known as fructose 6-phosphate (*left*).

DRAWING CURRENT FROM A STAR

The prospect of electricity made from sunlight—photovoltaic, or PV, energy as it is called—has tantalized scientists since 1839, when French physicist Edmond Becquerel showed that an electrode immersed in a conducting solution produces a voltage when exposed to light. Yet except in the space program, where PV cells serve as critical power sources in piloted and unpiloted missions alike, the use of photovoltaic energy has remained largely experimental.

Attempts to supply PV energy for Earth-based applications must first overcome a considerable cost barrier. In the early 1990s, the cost of generating one kilowatt-hour of energy by conventional methods—using fossil fuels such as coal and oil and fissionable materials such as uranium and plutonium—ranged from five to fifteen cents. Photovoltaic energy, by contrast, cost as much as twenty-five cents per kilowatt-hour. In part, the cost difference may be attributed simply to the fact that the technology and equipment for conventional methods is well established, whereas that for PV energy would require new investment.

By other measures, though, electricity supplied by the Sun has clear advantages over nonsolar sources. It does not pollute the environment with carbon monoxide or sulfur dioxide, which is responsible for the phenomenon known as acid rain, nor does it leave a residue of hazardous radioactive waste that requires storage and monitoring for thousands of years.

Through the agency of photovoltaic cells like the ones depicted below, sunlight has the potential to be a safe and reliable source of energy for all of Earth's inhabitants. But first, engineers must devise ways to store and distribute the Sun's energy to make it available 24 hours a day, 365 days a year to even the least sunny parts of the globe.

The photovoltaic panel shown here has a surface array of 480 circular cells, each of which collects sunlight and turns it into electricity *(box, opposite)*. Although the design of this PV generator calls for a stationary panel, other designs feature swiveling bases that allow panels to track the Sun across the sky. For greater efficiency, some designs use lenses to focus light onto the cells.

Solar Cells of Silicon

Most PV cells consist primarily of silicon, an element with four negatively charged outer, or bonding, electrons per atom. In pure silicon, the atoms form a lattice, each bonding to four others by means of their electrons. But to create the chemical instability needed to generate an electric current, two other elements are introduced—boron, with three bonding electrons per atom, and phosphorus, with five. A bit of boron is mixed in as the silicon crystals are grown. Because the boron atoms can create only three bonds in the silicon lattice, they leave a positively charged "hole" where the fourth bond would be. When one face of the silicon-boron cell is exposed to gaseous phosphorus, the phosphorus atoms form four bonds with the silicon, leaving the fifth phosphorus bonding electron unattached. With free negative electrons in one area of the cell and positive holes in the other, a current becomes possible (below).

Boron

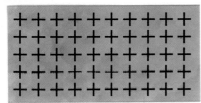

Phosphorus (n)

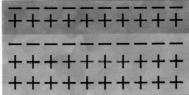

Boron (p)

Phosphorus (n)

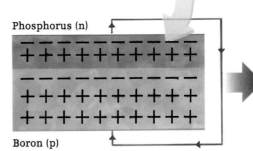

Boron (p)

A photovoltaic cell begins as a block of silicon into which boron is incorporated as the cell is being made. Boron, with three bonding electrons per atom to silicon's four, creates a wealth of so-called holes—positively charged areas that unattached electrons could fill. One face of the silicon block is then exposed to phosphorus gas.

The phosphorus permeates the silicon, forming a thin layer rich in free electrons. Some electrons promptly move across the junction between the negative (n) and positive (p) layers, even as some of the holes move in the opposite direction. This sets up a negatively charged barrier on the p side and a positively charged barrier on the n side.

When photons from the Sun strike the n layer of the PV cell, they knock loose electrons from the silicon atoms. Repelled by the surrounding free electrons in the n layer and unable to penetrate the barrier at the junction between layers, the electrons are easily shunted into an external circuit, producing an electric current.

ORBITAL HARVESTERS OF ENERGY

Future generations may harness the Sun to power their cities and transportation systems by building solar collectors in Earth orbit, far above the veil of the planet's atmosphere. A globe-girdling armada of hundreds of satellites, each with many square miles of solar cells, could convert sunlight to electricity virtually all day and all night, year round.

Circling above the equator at an altitude of 22,300 miles, the satellites would complete one revolution around the planet in twenty-four hours—the time it takes Earth to rotate once on its axis—thus remaining stationary relative to a given spot on the surface. This so-called geosynchronous orbit will enable the satellites to transmit their electricity to fixed receiving stations on land or sea.

The space-based generators could use either lasers or microwave transmitters to beam their power down to Earth. The more directed beam of visible-light laser transmission would permit the delivery of small packets of power to large numbers of receptors, but microwave transmission, because it is the better-understood technology, is more likely to be implemented. Since microwaves are much lower in energy than visible light, however, larger beams and ground-based antennas as much as six miles in diameter would be needed to transmit and harvest the energy efficiently. Complicating the efficiency issue are safety concerns that would require the beam to be further diffused at the source.

In any event, the cost and logistics of launching a fleet of satellites are daunting. To loft the necessary components from Earth and assemble them in space would consume trillions of dollars and hundreds of launches. However, NASA engineers estimate that transportation costs could be pared to one-tenth that amount if the components were built on the Moon. Not only are essential raw materials such as aluminum, iron, and titanium readily available, but sending components into space from the lunar surface would require an escape velocity of only 5,400 miles per hour, compared with the 25,000-mile-per-hour speed necessary to break free of Earth's gravity; the lower escape velocity would drastically reduce fuel costs.

4/A Ceaseless Wind

Shimmering rays of the aurora borealis, or northern lights, fan out from a central point—the direction of Earth's magnetic field at Fairbanks, Alaska. Touched off when charged particles of the solar wind excite atoms and molecules of oxygen and nitrogen in the atmosphere, the shape-changing lights occur nightly near both poles.

igh in the mountains of southern Mexico, on the crystal-clear morning of March 7, 1970, the small village of Miahuatlán de Porfirio Díaz seemed more the site of a love-in than of a scientific expedition. Solar enthusiasts of every description had gathered to witness one of those brief, mystical moments when the Sun disappears at midday. Under the somewhat baffled gaze of Miahuatlán's residents, a throng of transcendentalists, television reporters, and various tie-dyed travelers filled the town's main square, eagerly awaiting the forty-ninth—and, as it turned out, the most revealing—total solar eclipse of the twentieth century.

Apparently oblivious to the carnival atmosphere, several hundred solar astronomers feverishly adjusted their cameras, telescopes, and spectrographs in preparation for observing the Sun's normally invisible outer atmosphere, or corona, which would appear as a shimmering halo of pearl-gray light against the blackened sky. The scientists had journeyed from around the world to spend a few precious minutes in the dark at the one spot on Earth where totality—the time when the Moon completely masks the solar disk—would be the longest and the combination of altitude, weather, and sky clarity the most propitious for viewing. Solar experts would also study the event from various favorable locations along the hundred-mile-wide swath that the Moon's shadow would trace across Central and North America as far as Nova Scotia.

Among the stargazers at Miahuatlán was Donald Menzel, dean of solar scientists, former head of the Harvard College Observatory, and a legendary figure among eclipse watchers. In the course of nearly fifty years, between 1918 and 1966, Menzel had witnessed twelve totalities, beginning with his first as a Colorado teenager, in remote locales from Siberia to Saskatchewan. He and his associate, Harvard astronomer Jay Pasachoff, had made the trip to Mexico in the hope of learning more about the composition of the corona by recording the spectral signatures of chemical elements that are present in the coronal gas. Having worked together on other eclipses, the two had honed their observing procedures to a flawless routine. Yet despite their experience—and contrary to their expectations—they and their colleagues were about to see something altogether new.

At 11:35 a.m., as the last sliver of the Sun disappeared behind the black disk

During the total eclipse of March 7, 1970, the Sun displayed a uniform coronal halo in an ordinary photograph *(below)* and a much more complex structure in an image made with a radial density filter *(right)*. Darker at the center than at the edges, the filter dimmed the glare from the bright inner corona, revealing streamers at the Sun's poles and a gaping hole in the outer corona that vents the plasma of the solar wind.

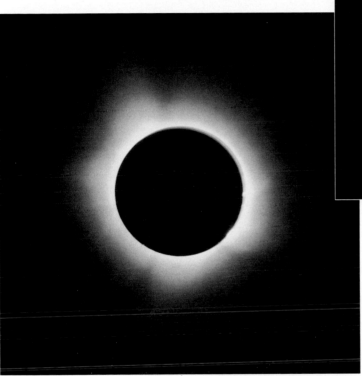

of the Moon, the corona suddenly and spectacularly burst into view, not as a smooth and seamless circle of light but as a patchy aura streaked with bright spikes and bulging streamers. The corona's ragged appearance was not in itself surprising: Irregularities had been documented before and were thought to correspond to disturbances on the Sun's surface. What was unusual this time was a sharp, distinct break—a dark gap—in the lower right portion of the otherwise glowing corona. Farther north along the path of totality, rocket-borne instruments tuned to x-ray and ultraviolet frequencies recorded the same hole.

The findings were tantalizing. After years of blackboard speculation, solar scientists finally had a convincing candidate for one of the most elusive pieces in the solar puzzle. Although the information would not be fully analyzed for years to come, this hole in the corona seemed to be an example of the long-sought escape valve for the powerful, pervasive, and mysterious phenomenon known as the solar wind.

THE DRAGON'S BREATH

Ancient myth spinners who described the Sun as a fire-spouting monster in the sky may not have been far off the mark. White-hot blasts of plasma—a rarefied mixture of electrons and protons resulting primarily from the ionization of hydrogen gas—spew out from the Sun like dragon's breath, moving at speeds as high as two million miles per hour and bathing planets, moons, asteroids, and comets in a seething wind that probably carries as far as the very edge of the Solar System.

Buffering Earth against this high-speed, high-energy torrent is the planet's magnetosphere—the surrounding region of space where magnetic forces generated by Earth hold sway, controlling the movement of charged particles. Roughly 40,000 miles from the planet's sunward side, the wind slams into the magnetosphere, compressing it into a giant bow-shaped shock front that deflects the wind around the globe, much as a boulder parts onrushing waters in a stream. On the planet's night side, the action of the wind pulls the magnetosphere out like taffy into a long, tapering tail that stretches well beyond the Moon's orbit. Similarly constrained and sculpted by the wind, the magnetospheres of other planets sport variations on this basic teardrop-shaped design.

Although most of the wind is diverted around Earth, billions of wind-driven electrons and protons still manage to breach the defenses. Pouring into Earth's atmosphere at the poles, they produce the dazzling light shows of the aurora borealis in the north and the aurora australis in the south. The wind also has more sinister effects of a wider purview, spawning geomagnetic storms that can disrupt radio communications worldwide, cause power surges in long-distance electric transmission lines, and perhaps, in ways that remain imperfectly understood, influence stratospheric chemistry, global warming, and long-term climatic change.

Of greater significance than these present-day influences is the pivotal role that the solar wind likely played in the genesis of the Solar System. Eons ago, as the early

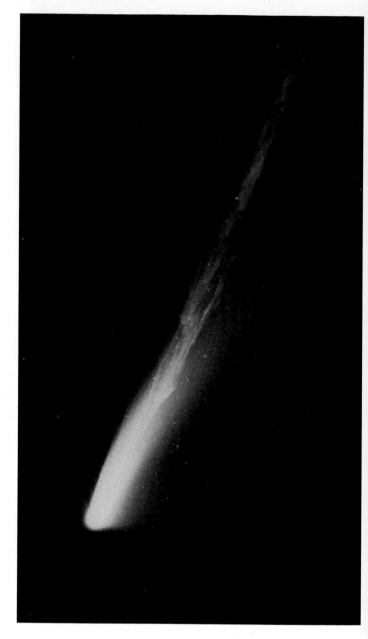

Sun began to spin into shape at the heart of a great amorphous cloud of dust and gas, a primordial solar wind may have blown away some of the surrounding disk of matter, thereby hastening the Sun's evolution from nebula to star and forever affecting the nature of the bodies in its orbit. The fact that the inner planets are rocky and the outer ones gaseous, for example, may in part be the result of the sweeping action of this wind, which carried lighter material farther out and left only larger, heavier chunks of matter closer in.

Although its effects are often clear, the solar wind itself is still not thoroughly understood. Scientists have been able to link sudden, massive bursts of wind to solar flares and other phenomena associated with the Sun's most active phases, but the continuous stream of plasma pouring from the Sun— even during its so-called quiet periods—has proved more difficult to explain. Its flow, for example, while unceasing, can vary wildly in velocity in a matter

Day-to-day changes in the long, straight plasma tail of comets—shown here in the significant thinning, over twenty-four hours, of the plasma tail of Comet Mrkos—alerted German astrophysicist Ludwig Biermann to the existence of the solar wind in 1951. Biermann determined that the plasma tail was moving faster than the shorter, curving dust tail, which scientists believed was bent by the weak pressure of sunlight. He then theorized that the plasma tail was responding to a stronger force: the constant bombardment by charged particles blowing from the Sun.

of hours, sometimes by as much as a million miles per hour. The exact source of this gusty wind, its long-term variations, and the force that propels it with such tremendous energy remain subjects of conjecture.

FIRST STIRRINGS

Despite its pervasiveness, the solar wind received little scientific attention until the 1950s. Before then, a few visionary astronomers had speculated that some force might emanate from the Sun, but they were at a loss to define it. In 1900, for example, Sir Oliver Lodge, a British pioneer in radio technology who also dabbled with the notion of spirit communication, suggested that the Sun was the source of a "torrent or flying cloud of charged atoms or ions" swirling through interplanetary space and giving rise to aurorae and magnetic storms on Earth.

Three decades later, another prominent scientist made the most of circumstantial evidence to posit the existence of some kind of mysterious radiation streaming from the Sun. German geophysicist Julius Bartels was one of the first scientists to employ statistical analysis to investigate cyclic phenomena, discovering among other things that the Moon's gravitational pull causes tides not only in the oceans but also in Earth's atmosphere. In 1932, when he applied the same methods to the investigation of magnetic storms on Earth, Bartels found that moderate storms seemed to recur at intervals of twenty-seven days, the same period as one full rotation of the Sun. Certain regions on the Sun, Bartels conjectured, must therefore be responsible for the magnetic disturbances on Earth. Because he could not link these disturbances with flares, sunspots, or any other observed activity on the surface, Bartels named the imagined areas M-regions, for "magnetic." The precise nature and location of these regions would remain undiscovered for well over three decades.

Other clues came from observations of one of the enduring oddities of interplanetary space—comets. Centuries ago, astronomers had noticed that the tails of comets always point away from the Sun. As a comet hurtles sunward on its orbit, its tail streams out behind it, but as it rounds the turn and heads back toward the depths of space, its tail precedes it. Most astronomers believed that the weak pressure of the Sun's light was pushing the comet tails away. Yet that theory ignored the fact that many comets wag two different tails: one the familiar curved fan of dust, and the other a longer and straighter tail of ionized gas, or plasma.

In 1951, Ludwig Biermann, then a professor of astrophysics at the Uni-

University of Chicago astrophysicist Eugene Parker points to a protuberance known as a helmet streamer in the Sun's outer corona, an area that he identified in 1958 as the point of origin for the solar wind. Parker's pioneering calculations showed that the outer corona expands continuously into space, bathing the Solar System in a hot wind of ionized gases.

versity of Göttingen in Germany, acknowledged that solar radiation pressure was strong enough to affect the comets' dust tails. But the comets' plasma tails, said Biermann, were a different story. They streamed away at much greater speeds and also showed more structure and variability, all of which suggested that they were sculpted by some force more potent than mere sunlight. Biermann proposed that streams of electrically charged particles—a component of something he called "corpuscular radiation"—poured out of the Sun at all times and in all directions to shape the plasma tails.

WINDS OF CHANGE

Biermann's view was widely accepted, yet it still did not explain the nature of these high-speed particles or the sort of mechanism that drove them. The first plausible theory would eventually emerge out of efforts to answer a seemingly unrelated question about the corona. Although the corona—as seen during a total eclipse—seemed to extend at most a few million miles from the solar surface, most scientists suspected that it stretched much farther. In fact, if the corona was as hot as the nearly one million degrees Kelvin that astrophysicists in the 1940s had calculated it to be, then it had enough energy to maintain its pressure and density even at great distances from the Sun. (Recent calculations have raised estimates of coronal temperature even higher, to almost one and a half million degrees.) In 1957, geophysicist Sydney Chapman demonstrated mathematically that the Sun's corona must extend at least beyond Earth's orbit.

In the end, it took a young astrophysicist named Eugene Parker to put two and two together. As a researcher at the University of Chicago in 1955, Parker had been captivated by Biermann's hypothesis. Chapman's theory, however, which was based on a so-called static corona permanently in place around the Sun as Earth's atmosphere is around Earth, seemed to disallow Biermann's high-speed flow of particles. After talking with both men, Parker concluded in a 1958 paper that Biermann's particles of corpuscular radiation could not penetrate the hot, charged gas of Chapman's extended corona—unless the corpuscular radiation and the extended corona were one and the same.

The proposition depended on a radical adjustment to Chapman's work, replacing the static corona with one that was continuously expanding. Parker theorized that the tremendous temperatures in the corona would cause it to be driven ever outward at speeds fast enough to allow it to escape the Sun's gravity and eventually spread into interstellar space. As the outer corona dispersed, it would be replenished by gases welling up from below.

Parker's model for this expansion and replacement described what he dubbed a solar wind. It began with an almost imperceptible expansion near the Sun, where the restraining force of gravity was the strongest. Because charged particles are excellent heat conductors, however, the wind would maintain its temperature and thus its energy as it expanded. The result, according to Parker, was that as the pull of solar gravity decreased with distance, the flow of coronal gas would steadily increase, reaching a velocity of more than 400,000 miles per hour at a point nearly two million miles above the solar surface. By the time this solar wind crossed Earth's orbit, it would be rushing outward at approximately 600,000 miles per hour, a speed that would account for the accelerations of comet tails associated with Biermann's corpuscular radiation. It was an attractive theory, but one based almost exclusively on intricate mathematical arguments, with little direct physical evidence to support it.

A challenge came almost immediately from astronomer Joseph Chamberlain, an associate of Parker's at the University of Chicago. Chamberlain argued in a series of papers that expansion by normal heat conduction would generate no more than a "solar breeze," with less than a tenth the velocity Parker had predicted. Was Parker's wind a gale or a whisper? Debate about the nature of the coronal expansion—indeed, about its very existence—was widespread, but for the moment it was also purely academic. Until instruments could be launched beyond the protective bounds of Earth's magnetic field, which effectively wards off incoming streams of charged particles, there was no way to confirm or deny Parker's hypothesis.

VERIFICATION FROM SPACE
The first tentative evidence began to arrive just before the end of the decade. On September 12, 1959, the Soviet Union launched the probe *Lunik 2,* which a few days later would crash-land on the Moon. Among the instruments aboard were charged-particle traps designed to provide data about the composition of interplanetary space. Similar collecting instruments were also carried by *Lunik 3* a month later and by the Soviet Venus probe *Venera 1* and

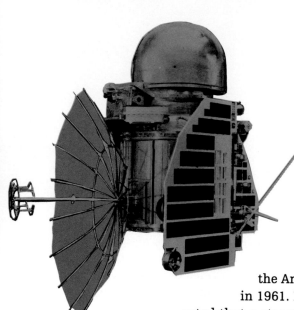

The fleet of space probes shown here yielded a number of insights about the solar wind during the 1960s. First aloft was the Soviet *Venera 1* probe *(left)*, which entered orbit on February 12, 1961. Charged-particle traps aboard the craft detected a continuous flow of plasma, or ionized gases, pouring from the Sun.

the American satellite *Explorer 10,* both launched in 1961. Measurements by all these spacecraft indicated that a stream of particles was rushing past Earth from the direction of the Sun at more than 500,000 miles per hour. Although the results seemed to support Parker's vision of solar outflow as a strong, persistent wind rather than a soft breeze, none of the Soviet vehicles made very long observations, and the American orbiter operated so near Earth that its information was considered suspect, possibly skewed by the planet's own magnetic field.

Then, in 1962, any doubts about the strength of the solar wind were removed by NASA's *Mariner 2* probe to Venus. One of the craft's instruments was a positive-ion spectrometer, a device that records the energies of individual charged particles as voltages that can be converted into velocity values. During nearly four months of observations, *Mariner 2* detected an almost continuous wind, ranging in speed from about 650,000 to more than 1,800,000 miles per hour. Even more remarkable, high-speed gusts recurred almost exactly twenty-seven days apart, and each one could be correlated with a period of heightened disturbances in Earth's magnetic field. Not only had Parker's theory about wind velocity near Earth been confirmed but also a link had been established between the wind and geomagnetic activity. Still, questions remained as to the driving force behind both the steady wind and its more blustery outpourings.

Investigations into another element of Parker's solar wind theory yielded further clues. Parker reasoned that since the wind was composed of charged particles, it would pull weak portions of the solar magnetic field along with it as it streamed into space, thus creating an interplanetary magnetic field with lines of force originating in the Sun. The effect, according to Harvard astronomer Robert Noyes, is akin to that of "running a comb through one's hair: The individual hairs are made parallel to the direction of motion of the comb." The analogy is complicated, however, by the Sun's rotation. Working from his estimates of solar wind speed, Parker calculated that the wind would take about four and a half days to reach Earth, during

The American *Explorer 10* satellite transmitted data for only sixty hours after its launch on March 25, 1961, but its plasma probe clocked the solar wind at an average speed of 175 miles per second and gauged its temperature in the range of 100,000 to 1,000,000 degrees Kelvin. Later measurements showed the wind's average velocity to be as high as 310 miles per second.

Orbiting Earth in a path that took it halfway to the Moon, the Interplanetary Monitoring Platform *(right)* revealed in 1963 that the solar wind sculpts the planet's magnetosphere, or magnetic envelope, creating a bow shock on the sunward side and a cometlike magnetotail on the other.

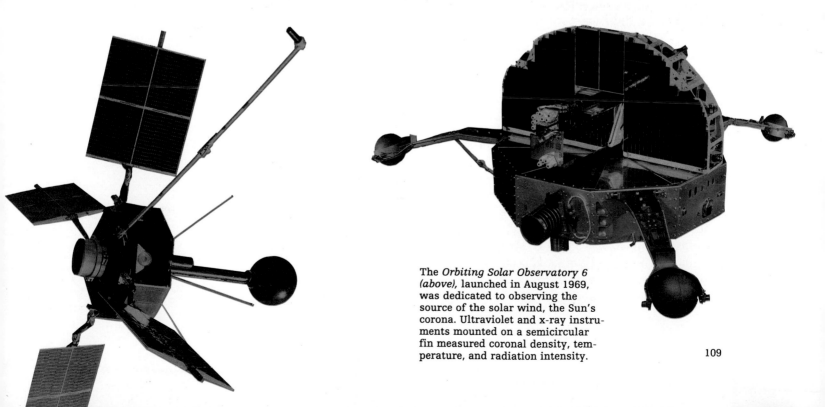

which time the Sun rotates about sixty degrees. With their feet firmly planted in the rotating solar surface, the magnetic field lines would thus be drawn into a giant spiral similar to the spray streams thrown out by the arms of a rotating lawn sprinkler.

Confirmation of an interplanetary field came in 1963 from the Interplanetary Monitoring Platform *(IMP-1),* a NASA satellite equipped for detecting magnetic fields and measuring such characteristics as their strength and the orientation of their lines of force. At the apogee, or farthest point, of its orbit, *IMP-1* was about 120,000 miles from Earth, well outside the planet's own magnetic field. From this vantage, its instruments not only registered the presence of an interplanetary magnetic field but were able to verify the field's spiral shape as well.

When *IMP-1* scientists Norman Ness and John Wilcox examined the satellite's data more closely, they realized that the curved magnetic field lines were organized into half a dozen large-scale, long-lived structures resembling the blades of a child's pinwheel. Ness and Wilcox dubbed the blades interplanetary magnetic sectors, because each one displayed a distinct magnetic polarity. That is, in a region of positive polarity, the field was said to point away from the Sun, while in a negative region, it pointed toward the Sun. (Direction in a magnetic field indicates how, for example, a simple compass placed in the field would orient itself, with its needle pointing along a field line either away from or toward the source of the field.) In addition, the polarity of each blade matched the polarity of a solar region from which the blade had apparently originated. The *IMP-1* data also revealed that alternating sectors of differing polarity washed over Earth every few days, a fact that was confirmed by observed variations in geomagnetic disturbances. And

The primary mission of NASA's *Mariner 2* spacecraft *(above)* was to observe the atmosphere of Venus. However, the probe also made more than 40,000 solar wind measurements along the way. Collecting 720,000 bits of data each day for a four-month period beginning on August 27, 1962, *Mariner 2* was the first instrument to detect gusts in the solar wind.

The *Orbiting Solar Observatory 6 (above),* launched in August 1969, was dedicated to observing the source of the solar wind, the Sun's corona. Ultraviolet and x-ray instruments mounted on a semicircular fin measured coronal density, temperature, and radiation intensity.

because each sector exhibited distinctive features, such as a particular pattern of variation in the field's intensity, the satellite was able to show that any one sector enveloped Earth once every twenty-seven days, in co-rotation with (although lagging behind) the Sun.

Coming on the heels of the *Mariner 2* findings about solar gusts at twenty-seven-day intervals, this discovery bolstered the conviction of many that there was a connection between specific regions of the Sun and the interplanetary phenomenon of the solar wind. "Evidence was building up," observed Noyes in 1982, "that somehow the Sun was unleashing its wind gusts from localized sources that rotated with the Sun once every 27 days." But the origin of the wind remained as elusive as it was in 1932, when Bartels introduced the idea of the mysterious M-regions: In the early 1960s, as in Bartels's day, no visible features on the Sun could be linked to either the magnetic disturbances on Earth or the newly discovered solar wind.

In the ensuing half-dozen years, valuable new data continued to pour in from a growing fleet of satellites. Increasingly sophisticated instruments were making it possible to study the Sun's electromagnetic radiation in a wide range of frequencies blocked by Earth's atmospheric veil—specifically, x-rays, gamma rays, and ultraviolet light. But the significance of some of the data did not immediately register with solar scientists. In 1968, for example, ultraviolet detectors aboard one of the satellites in the new Orbiting Solar Observatory (OSO) series discerned large areas in the corona that were distinctly fainter, radiating at much lower intensity levels than the normal corona and thus indicating that coronal material was more tenuous in these regions. Other ultraviolet and x-ray data revealed so-called coronal vacancies, where there was little or no radiation. Most scientists, however, convinced that the only good place to look for answers was in active regions, ignored these new findings. It would take the eclipse of 1970 to open their eyes.

DARKNESS THAT SHED LIGHT

With 50 million people living either within or no more than a day's drive from the path of totality, the March 1970 eclipse was guaranteed to attract a great deal of popular attention, but the scientific interest it generated was also especially intense. By a lucky quirk of celestial mechanics, the eclipse path passed directly over NASA's research launch facility on Wallops Island, Virginia, providing an ideal opportunity for detailed study. Thirty-four sounding rockets—small, suborbital vehicles carrying scientific experiments —were launched from Wallops Island alone, twenty-five on the day of the eclipse, the others in the few days before and after. Another fourteen flew above Elgin Air Force Base in Florida, while two more took off from White Sands Missile Range in New Mexico. Astronomers sent still other instruments aloft in high-flying airplanes that chased along the eclipse path to extend totality by several minutes. In space, six satellites—*OSO-5* and *OSO-6*, the Applications Technology Satellite, the joint U.S. and Canadian satellites

WHY THE SUN WOBBLES

Although it is typically depicted as a stable center around which the nine planets revolve, the Sun is in fact a moving participant in the Solar System's orbital dance. According to Isaac Newton's law of universal gravitation, when one body orbits another, both actually circle their common center of mass, a point known as the barycenter (from the Greek word for "heavy"). As demonstrated below, the barycenter for any two celestial bodies is always located closer to the more massive object. The greater the disparity be-

tween the two masses, the closer the barycenter will be to the more massive object and thus the smaller that object's orbit around it.

Given the Sun's virtual monopoly on Solar System mass—99.9 percent of the total—the barycenter for the entire system deviates only slightly from the Sun's own center, so that its orbital motion is no more than a wobble. The following pages illustrate some of the complexities in this wobble. They arise from the continuously shifting relationships between the Sun and its planets—intricacies that to a distant observer unable to see the planets themselves would be a tip-off to their presence.

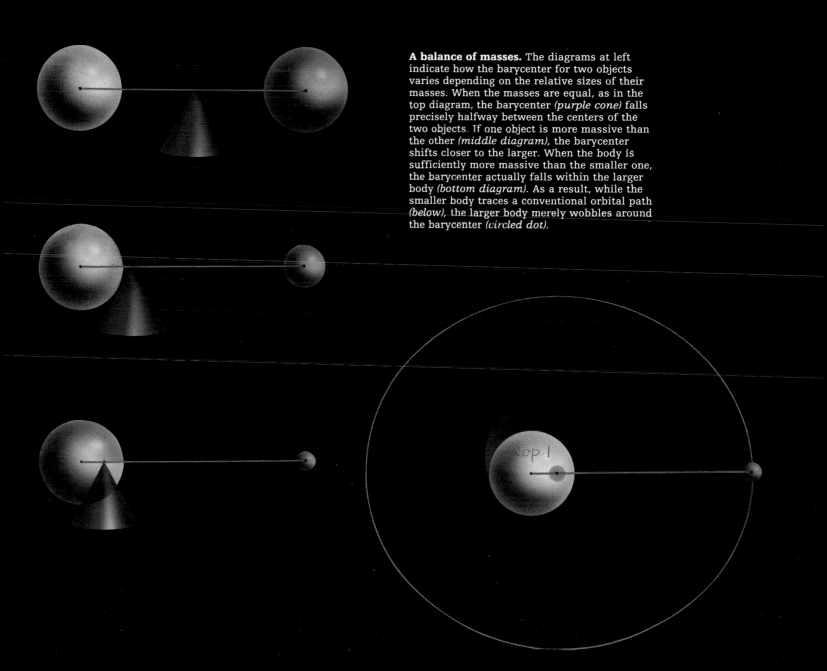

A balance of masses. The diagrams at left indicate how the barycenter for two objects varies depending on the relative sizes of their masses. When the masses are equal, as in the top diagram, the barycenter *(purple cone)* falls precisely halfway between the centers of the two objects. If one object is more massive than the other *(middle diagram)*, the barycenter shifts closer to the larger. When the body is sufficiently more massive than the smaller one, the barycenter actually falls within the larger body *(bottom diagram)*. As a result, while the smaller body traces a conventional orbital path *(below)*, the larger body merely wobbles around the barycenter *(circled dot)*.

A Planetary Tug of War

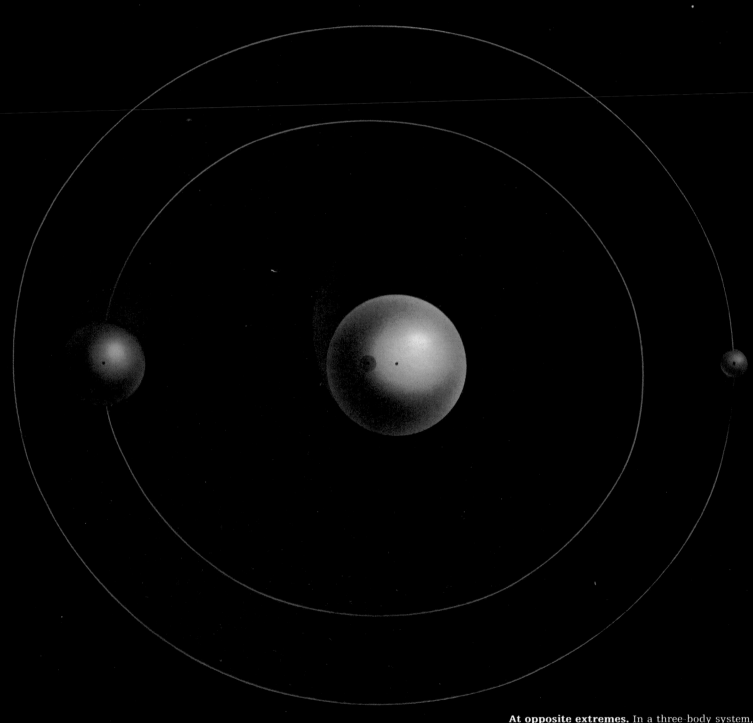

At opposite extremes. In a three-body system, when the two planets are on opposite sides of the star, the barycenter *(circled dot)* is as close as possible to the star's center *(dot)*. The more massive planet of the two pulls the barycenter to the left, but not as far as it would if there were no planet on the other side to counterbalance the effect.

If it had only one planet, the Sun would follow an unwavering orbit around the barycenter, because the barycenter itself would maintain the position dictated by the ratio between the two masses. But when more than two bodies are involved, as in the Solar System, the relationship between masses becomes much more complex, causing the barycenter to shift and the Sun's orbit to fluctuate.

Basic orbital mechanics accounts for the change in balance. Planets nearer the Sun, which must orbit more swiftly to offset the Sun's stronger gravitational pull upon them, are always catching up with and passing planets farther out. This continuously alters the distribution of the combined mass of all the planets, sometimes apportioning it fairly evenly around the Sun and sometimes concentrating it on one side. The three-body system diagramed below illustrates in simplified form how this shifting distribution of planetary mass works to pull the Solar System's barycenter closer to and then farther away from the Sun's center, which in turn causes the Sun's wobbling orbit to shrink and expand.

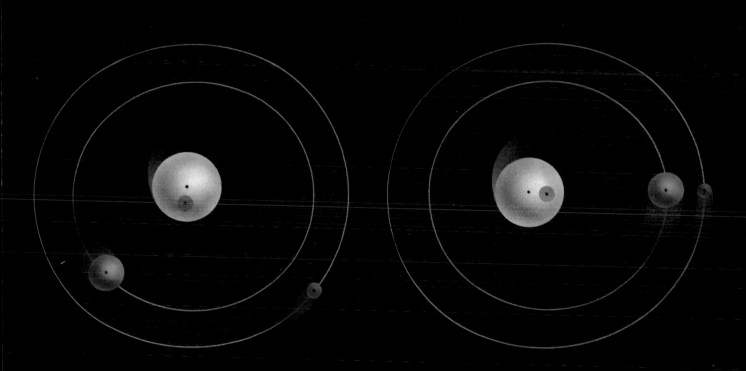

Drawing closer together. Because the inner planet travels faster and over a shorter distance, it begins a second orbit before the outer planet completes its first. With both planets now on the same side of the star, their masses start to pull together on the barycenter, shifting it slightly farther from the star's center.

Planets aligned. Eventually, the inner planet catches up with the outer planet so that they line up on the same side of the star. With both planetary masses acting as one, the barycenter is drawn to its maximum distance from the center of the star. It will then start to shift back toward the star's center as the planets move out of alignment.

CLUES IN A SHIFTING SPECTRUM

The orbital behavior of the Sun or of any star is far more than a curiosity of physics. Because planets around a distant star would be too small and faint to be observed directly from Earth, astronomers rely instead on detecting the wobble such planets would cause. The process begins with an analysis of starlight using a spectrograph, which splits white light into the full range of visible frequencies, from the longer waves of red to the shorter waves of blue and violet. Depending on its composition, a star exhibits across

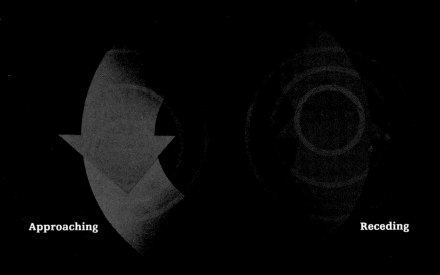

Approaching

Receding

A wavelength shift. As shown at left, light waves from an object compress toward the blue end of the spectrum as the object approaches an observer but stretch out toward the red as it recedes. The three spectra below indicate how a wobbling star's spectral lines would move from their expected position *(top)* toward blue *(middle)* and red *(bottom)*. Shifts have been exaggerated for clarity.

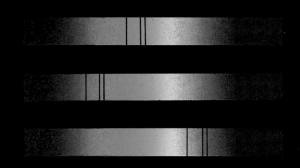

Observer

Plotting wobbles. The curves below plot a star's orbit in terms of radial velocity—its receding *(red)* or approaching *(blue)* speed. With only one planet *(left)*, the star shows a regular periodic motion as it orbits smoothly around the barycenter. But two planets distort the wobble *(below)*, maximizing velocity when the planets come together and minimizing it when they are far apart.

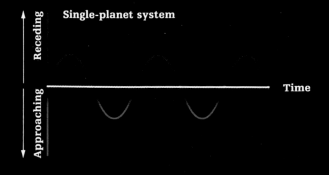

Single-planet system

Receding

Approaching

Time

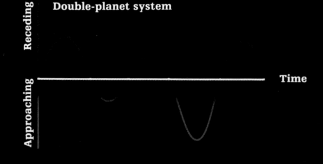

Double-planet system

Receding

Approaching

Time

its spectrum a characteristic pattern of dark bands called spectral lines, produced when different chemical elements absorb different wavelengths of light.

Once a star's spectral line pattern has been identified, planet hunters can look for a wobble. According to the Doppler effect—the phenomenon that makes a whistle sound higher as a train approaches and lower as it recedes—a star's spectral lines will shift toward the blue end of the spectrum as the star moves closer to the observer and toward the red as it moves away.

Such shifts would indicate that the star is in orbit around a barycenter created by the presence of either an unseen stellar companion or one or more planets.

Plotting spectral shifts may also reveal how complex a system is. While a one-planet system would create a steadily repeating alternation between red shift and blue shift, the Sun's own spectrum observed from afar would generate a highly irregular pattern, undoubtedly leading alien astronomers to hypothesize the existence of multiple planets.

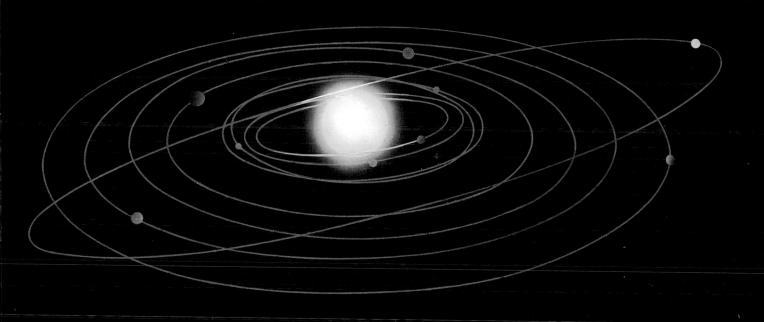

A distinctive pattern. The complex effects of a nine-planet system *(above)* become apparent in a hypothetical graph *(below)* of the Sun's radial velocity over an eighty-four-year period as it might be measured by a distant observer. Variations in the curve reflect the degree of the Sun's wobble as the barycenter moves in response to shifting alignments of the planets.

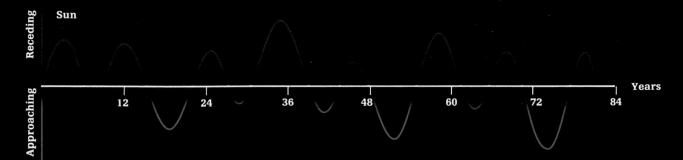

Sun

Receding

Approaching

Years

12   24   36   48   60   72   84

Astronauts aboard Skylab, the U.S. space station, studied the Sun during the course of three separate missions, capturing more than 150,000 images between May 25, 1973, and February 8, 1974, when the last crew departed. The craft's Apollo Telescope Mount—visible at the center of the four solar panels—held eight monitors *(below)*, including five that analyzed x-ray and ultraviolet radiation, wavelengths difficult or impossible to detect from Earth.

Braced by a cross-shaped optical bench, the instruments housed in Skylab's Apollo Telescope Mount were *(clockwise from top)* an x-ray telescope *(white)*, a second x-ray telescope, an ultraviolet (UV) spectrograph, a hydrogen-alpha telescope, an extreme-UV spectroheliograph, a UV spectroheliometer, another hydrogen-alpha telescope, and a coronagraph.

Alouette 1 and *Alouette 2,* and the International Satellite for Ionospheric Studies—all turned their detectors toward the darkened Sun. On the ground, meanwhile, hundreds of solar experts recorded a host of coronal phenomena not visible at any other time.

Occurring as it did almost at the peak of the Sun's activity cycle, the eclipse revealed a corona of exceptional brightness and intricate structure. The most interesting feature, of course, was the sharp gap yawning in the southern half of the corona. The visual display was compelling, but the crucial detail would come from magnetic readings taken by the sounding rockets. Typically, the lines of force of a solar magnetic field form a closed loop, which rises into the corona from a region of positive polarity and then turns back into a region of negative polarity on the solar surface. Within the newly discovered coronal hole, however, the localized magnetic field was unipolar, its lines of force bursting straight out from the Sun.

Later that year, the discovery of such a magnetic configuration was corroborated by solar physicist Allen Krieger and his colleagues at American Science and Engineering, an independent scientific research firm in Cambridge, Massachusetts. Krieger and his team used rocket-gathered x-ray images of the solar disk to pinpoint a "magnetically open structure" in the solar corona. Then, employing a complex mathematical formula that allowed them to extrapolate the solar wind's behavior from satellite observations, they were able to trace the path of a stream of high-speed particles back from Earth to this very same open structure on the Sun. The implication was clear: Coronal material was somehow escaping and blowing through interplanetary space, dragging field lines with it. The stage was now set for the most intense solar observation program ever mounted—the Skylab experiments, which would uncover definitive evidence of a link between the Sun's pockmarked corona and its high-speed solar exhaust.

A SOLAR LAB IN SPACE

Weighing in at more than 100 tons, Skylab was at the time of its launch in May 1973 the United States' largest space vehicle—and the most complex. Three separate crews of astronauts were scheduled to rendezvous with the boxcar-size laboratory, each to conduct low-gravity experiments, make remote surveys of Earth resources, and observe the Sun with a sophisticated battery of telescopes and cameras.

The technological centerpiece of the craft was the outboard Apollo Telescope Mount, or ATM. Bristling with eight separate instruments *(opposite),* the ATM was designed exclusively for solar scrutiny. Together, the telescopes could examine the Sun in the full range of wavelengths from x-ray through ultraviolet to optical light. Access to the higher-frequency radiation was possible only in space, but even visible-light instruments benefited from Skylab's lofty perch 270 miles above Earth. The craft's coronagraph, for example, which created artificial eclipses by blocking the solar disk with a small central mask, could view the surrounding corona against a sky as

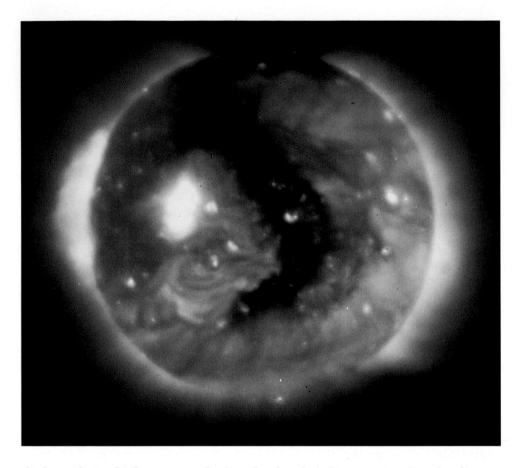

A false-color x-ray photograph taken from Skylab on August 23, 1973, reveals a gaping hole *(black)* in the Sun's outer corona. The bright white spots denote regions of intense magnetic activity in the upper reaches of the corona; the faint swirls are magnetic loops holding the corona in place. No magnetic loops traverse the coronal hole; instead, the Sun's magnetic field lines stretch into interplanetary space. Gases welling up from beneath the corona flow out along these lines, creating high-speed streams of solar wind.

dark as that which appears during the few brief moments of real eclipses.

Never before had the Sun been so closely studied. In the eight months that Skylab was in operation, the astronauts recorded and stored data from the various ATM telescopes on nearly thirty canisters of photographic film containing some 163,000 exposures. The coronagraph alone logged a total of nearly nine months of observing—in contrast to the eighty hours of coronal observations made from the ground during all the natural eclipses since the invention of photography in 1839. The caliber of the images was much higher than had been anticipated, a result of the crews' unexpected talent for taking the right pictures at the right time. In fact, space scientists who had worried that human operators might be less reliable than automatic instruments changed their tune upon viewing the returned solar films. Almost every solar event of any significance—whether predicted or totally unanticipated—had been captured.

An additional boon was that astronaut Edward Gibson, a member of the third crew, had a doctorate in engineering physics and before his spacefaring days had written a textbook on solar phenomena. (Ironically, it was entitled *The Quiet Sun*—precisely the opposite of the Sun's state during Gibson's eighty-four-day shift.) Like the two crews before them, Gibson and his fellow

Skylab astronauts Gerald Carr and William Pogue saw the Sun oscillate dramatically between activity and quiescence.

Skylab's lengthy observing run yielded a number of insights. Naturally, the most significant to researchers hungry for more information about the solar wind related to coronal holes. X-ray images, in which areas of less intense radiation stood out as great dark blotches, proved particularly helpful. While the Skylab astronauts watched the blotches rotate with the Sun, earthbound observers using the same extrapolation technique employed by Krieger in 1970 followed back to the Sun the trail of high-speed streams of charged particles showering Earth. Each particle path led to a spot where the Skylab astronauts had observed the position of a coronal hole four and a half days earlier—precisely the transit time of the solar wind. A direct relationship also began to emerge between the size of a coronal hole and the velocity of the wind, with the largest holes emitting the speediest gusts. There could no longer be any doubt that coronal holes were the source of the solar wind.

Data from the ATM telescopes also indicated that these apparent voids in the corona were not completely empty; low levels of both x-ray and ultraviolet emissions suggested that small amounts of hot material remained. Measurements of this residual radiation—made by satellites and rocket-borne detectors as well as by Skylab—indicated that the density of a typical hole is a mere 10 percent that of the normal corona, and the temperature half a million degrees cooler. Although astronomers had no trouble attributing the lower density and temperature to the drop in pressure that results when coronal mass is blown away, they were still in the dark as to why the solar energy that usually heats the corona should instead in these regions be impelling it into space.

THE SKYLAB LEGACY

The last of Skylab's three scheduled crews departed on February 8, 1974, leaving the giant spacecraft to orbit unmanned and unseeing. Although NASA had no plans for further missions, solar scientists hoped that eventually the laboratory would be put back into service. But the Sun itself thwarted their wishes. Late in 1978, solar flares associated with rising sunspot activity bombarded Earth with high-energy particles, heating the atmosphere's outer fringe and causing it to expand farther into space. The resulting increase in drag pulled Skylab into an uncontrollable—and fatal—spiral toward Earth. On July 11, 1979, the craft splintered apart during reentry over the Indian Ocean, its bits and pieces falling into the ocean and across the Australian Outback in a fiery metallic rain.

Despite its ignoble end, Skylab left a distinguished legacy to solar science. It had so improved the understanding of coronal holes, for example, that earthbound astronomers were able to devise a subtle technique for detecting them from the ground. Aware that helium moving from the solar surface into the chromosphere is partly excited by radiation from abutting regions of the corona, solar scientists employed a spectrograph tuned to the spectral lines of helium to measure the shape and size of coronal holes. Continuous monitoring of coronal holes from solar observatories at Arizona's Kitt Peak and New Mexico's Sacramento Peak has revealed that most holes are neither constant nor permanent: Like other features on the Sun, they come to life, evolve, and eventually die away in periods ranging from a few weeks to a few months. The observations have further disclosed that when a large coronal hole appears near the solar equator, a high-speed solar wind of more than 1.5 million miles per hour is often recorded by satellites in Earth orbit. This finding confirmed the link between exceptionally large coronal holes and especially high speed winds.

THE MYSTERIES OF MAGNETISM

By the mid-1970s, a clearer picture was emerging of the role played by the Sun's magnetic field in the dynamics of both the corona and the solar wind —a picture that would help explain the likely origin of the often overlooked steady wind first hypothesized by Parker and later proved by observation. Solar specialists recognized that, except where there are holes, the Sun's magnetic field confines the gases of the corona, holding the corona in place and giving it shape and structure. The magnetic field lines that anchor the corona originate deep in the solar interior, then break through the surface and soar as high as 250,000 miles into the corona before plunging back to the solar surface at a different spot to create a coronal scaffolding of huge bipolar arches.

Within coronal holes, of course, magnetic field lines stretch far beyond the corona, allowing material to escape and thus engendering high-speed gusts of solar wind. But scientists were now finding that open field lines also existed right next to certain closed loops, where they seemed to give rise not to strong blasts but to a constant, slow leaking of coronal material—Parker's steady wind. As Randolph Levine, a Harvard astronomer who helped interpret data from Skylab's ultraviolet telescope, pointed out, "The location of at least some sources of the wind should be near where the highest closed loops are rooted. Such places might look like the part in a head of hair, with the open lines sticking out like a persistent cowlick." In order for such field lines to make their escape, Levine conjectured, they must be driven past a critical threshold—about 250,000 miles above the surface—beyond which their magnetic force is too weak to contain coronal gas.

To this day, however, solar scientists have not agreed on the mechanism that opens magnetic field lines, whether in coronal holes or near closed loops. One theory, first proposed in the early 1970s by astronomer Joseph Hollweg of the University of New Hampshire, relates the phenomenon to the action of so-called magnetohydrodynamic waves, a form of plasma wave that propagates along magnetic field lines, perhaps as a result of turbulent fluid motions beneath the solar surface. As they ripple through the corona, these waves may push plasma outward in a kind of shock front, accelerating charged particles with enough force to break magnetic loops holding the corona in place.

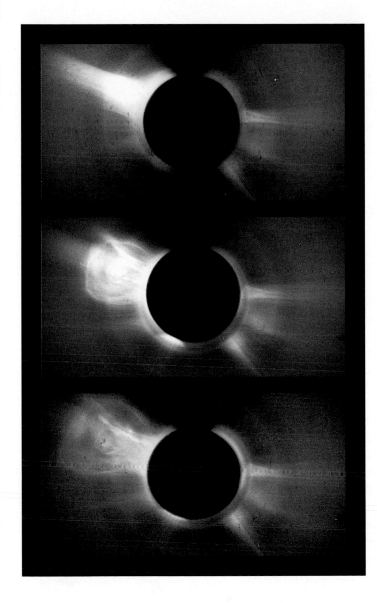

In the course of two hours, the Sun displays one of its more violent phenomena—a coronal mass ejection. At top, the corona displays wispy streamers, marking regions where the solar wind escapes relatively quietly along magnetic field lines. Expanding coronal gas stretches the field lines ever further *(middle)* until they are ready to snap *(bottom)*, spewing billions of tons of charged gas particles into space at the rate of 300 miles per second. The dark blotch at the top of each frame is the shadow of an occulting disk, a mask to make the corona visible.

Even as this and other hypotheses were being investigated, new information was emerging about another means by which material exits the Sun. Back in 1973, the white-light coronagraph aboard Skylab had observed a great gout of plasma rising like a prominence above the edge of the solar disk. Climbing higher and higher, it reached speeds of almost a million and a half miles per hour. When it finally burst free of the Sun's magnetic field, the loop-shaped outpouring released an awesome spray of gas. It was the first of more than a hundred similar—if not quite so spectacular—eruptions that the Skylab astronauts would witness.

Between 1980 and 1984, coronagraphs aboard both NASA's Solar Maximum Mission and the U.S. Air Force's *P78-1* satellite recorded scores more of the so-called coronal mass ejections (CMEs). The *P78-1* observations suggested that these eruptions take the shape of great bubbles blown out of a tiny, tubelike hole in the corona. One or two CMEs balloon out of the corona daily, each carrying 10 billion metric tons of material, a significant fraction of what the solar wind disperses in a day. More might be known about coronal mass ejections had the Air Force not decided to shoot down the *P78-1* satellite during a test of an antisatellite rocket in 1985.

WHITHER THE WIND?

Although most studies of the solar wind look inward to its source, scientists are also intrigued by the eventual fate of the outflowing plasma stream. Before the 1980s, astronomers predicted that somewhere beyond the orbit of Jupiter the weakened wind would collide in a shock front with the tenuous gas thought to fill interstellar space, and its magnetic field lines would be deflected to reconnect with the Sun. By June of 1988, however, the *Pioneer 10* spacecraft launched in 1972 had traveled six times farther than the expected cutoff point near Jupiter, and still it could feel the solar wind flowing by. Scientists hope that *Pioneer 10*—or either of the two Voyager spacecraft also heading out of the Solar System—will eventually allow them to chart the wind's full extent, although they are increasingly prepared for the possibility that, rather than ending in a collision, the wind may simply peter out, mingling gently with other gases between the stars.

Solar astronomers are also seeking to expand their knowledge of the solar wind beyond the narrow, essentially two-dimensional region of the Sun's ecliptic plane, where most of the planets reside and where, until now, all observations have been conducted. Late in 1990, the Ulysses spacecraft, a joint venture of NASA and the European Space Agency (ESA), will begin an epic journey. After being placed in Earth orbit by the space shuttle, Ulysses will be boosted out to rendezvous with Jupiter fourteen months later. With Jupiter's powerful gravitational field acting as a giant slingshot, the spacecraft will be flung out of the ecliptic plane on a trajectory that two and a half years later will take it first over the Sun's southern pole, then up and over the northern pole. Three and a half years after launch, the spacecraft will begin sending its data back to Earth, giving researchers their first truly three-dimensional view of the Sun and the vast volume of space that comes under its influence.

AN INTERNATIONAL ARMADA

The venturesome Ulysses will eventually be joined in space by a flotilla of spacecraft launched under the aegis of the International Solar Terrestrial Physics program (ISTP), a campaign drawing on the resources of American, European, and Japanese space agencies and designed to investigate the Sun's remaining mysteries. Among ISTP's scientific spacecraft will be four satellites collectively known as Cluster, to be placed in near-Earth orbit by ESA's Ariane rockets. Carrying several joint European-U.S. experiments, the quartet will fly in formation, separated by distances of a few hundred to a few thousand miles, and record small-scale phenomena produced when the solar wind slams into Earth's magnetic field.

Two NASA spacecraft designed to work in tandem are also to be launched by the middle of the decade. The first, called Wind, will be positioned permanently on Earth's sunlit side to monitor shocks generated by the incoming wind; the second, Polar, will fly over Earth's own poles to watch any resultant auroral activity. At the same time, Geotail, an American-Japanese satellite, will travel in a long, looping orbit that will take it through the magnetosphere's cometlike tail.

The premier spacecraft in this international effort will be the Solar and Heliospheric Observatory (SOHO), being built by ESA, again in cooperation with NASA. Involving more than a score of American and European universities and research centers—and literally hundreds of individual scientists—SOHO will attempt a definitive analysis of the processes that form, heat, and maintain the Sun's corona and give rise to the solar wind. Scheduled for launch in 1995, SOHO will eventually reach a stable position at one of the places between Earth and the Sun, known as libration points, where the gravitational pull of both bodies is nearly equal. One million miles beyond the magnetosphere on Earth's daylight side, SOHO will be able to observe the Sun, its corona, and its outpouring solar wind continuously—without interference from either Earth's atmosphere or its day-night cycle. Long accustomed to having to extrapolate details of the wind's behavior, solar scientists eagerly await this first-ever opportunity to spy on the solar wind from the time it departs the corona until it envelops Earth.

THE POWER OF A SOLAR GALE

nterplanetary space, once thought to be an utter void visited only by the occasional subatomic particle from beyond the Solar System, is in fact a maelstrom of charged particles, most of them bits and pieces of the Sun itself. Traveling at one million miles per hour, this electrified gas, or plasma—made up of more or less equal numbers of hydrogen nuclei and free electrons, along with a small percentage of helium nuclei and other ions—is the stuff of the solar wind.

Pouring out of the Sun at the prodigious rate of one million tons per second, the plasma gale buffets the Sun's clutch of planets and extends four to eight billion miles beyond the orbit of Pluto. But it is exceedingly tenuous: By the time it swirls past Earth, it averages one million-trillionth the density of Earth's atmosphere at sea level.

Rarefied as it is, the solar wind—like even the mildest earthly breeze—possesses the power to bend and move things in its path. In fact, astronomers first inferred its existence from the way the tail of a comet always flows away from the Sun, no matter where the comet is in its orbit. The wind's true strength only becomes manifest, however, in the realm of electric and magnetic forces. As shown on the following pages, the solar plasma carries within it the magnetic field generated by the Sun itself. As the magnetized wind courses around the planets, many of which are wrapped in their own magnetic sheathing, the cosmic friction generates stupendous electric currents—source of raging magnetic storms, dancing aurorae, and approximately enough energy to supply in one day France's electrical needs for six months.

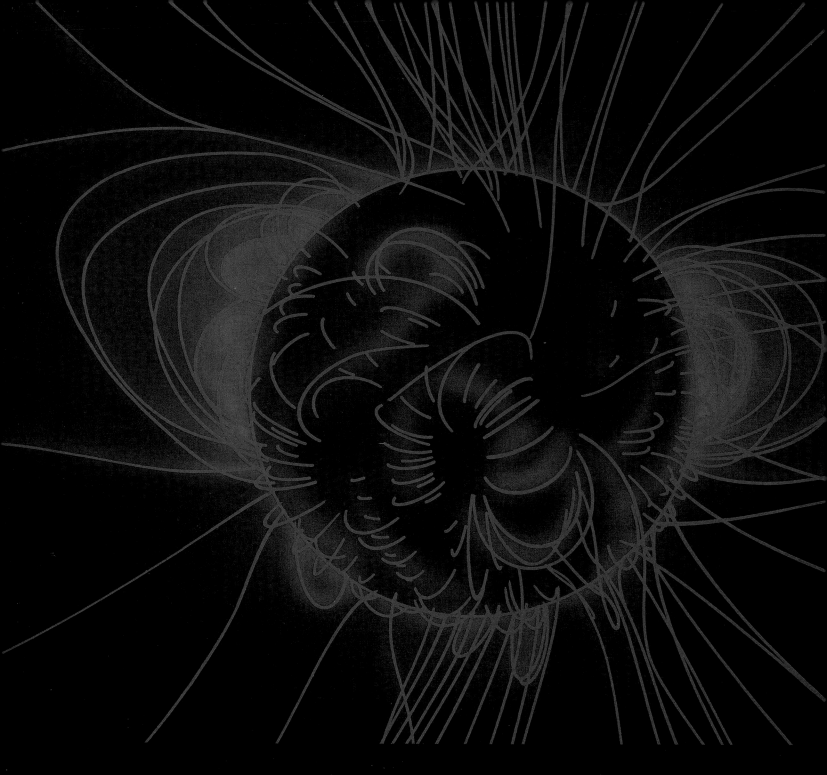

BREAKING OUT OF MAGNETIC BONDAGE

In the Sun's corona, two great forces are locked in constant battle. The lines of the Sun's magnetic field, running generally north-south but spun into twisted, arcing strands by the star's rotation, struggle to confine the coronal plasma, a writhing shroud of superheated charged particles. (For reasons scientists still do not fully understand, the corona, at temperatures between 1.5 and 2 million degrees Kelvin, is some 200 times hotter even than the estimated temperature of the surface.)

The relative strength of the opponents waxes and wanes with distance from the Sun's surface. Within 250,000 miles, the field lines contain enough magnetic force to keep the coronal plasma in check. Though pressured by the expanding gas, the lines maintain two footholds in the Sun—one negative pole, one positive—causing the plasma's charged particles to follow the vaulting paths of magnetism out of, and then back into, the surface.

Beginning about half a solar radius above the surface, however, the forces of magnetism weaken. The

High-speed solar wind. The wind that issues from coronal holes escapes along unipolar magnetic field lines that fly into space, with only one end attached to the Sun. The most tenuous wind rushes from the hole's center, sometimes reaching speeds of two million miles per hour. No wind emanates from the closed field lines flanking the hole, where trapped plasma is ten times more dense.

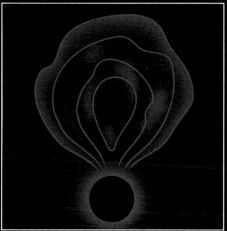

Explosive solar wind. Encased and threaded by loops of magnetism, bubbles of superheated coronal gas hurtle out of the Sun at speeds ranging from 22,000 to more than two million miles per hour. Known as coronal mass ejections, these eruptions result from huge coronal disturbances, occur on the average of twice daily, and spew 10 to 100 billion tons of solar material into space.

Quiet solar wind. Traveling at an average of 700,000 miles per hour, the solar wind that emanates as so-called helmet streamers is deemed "quiet" by solar scientists. The streamers form when open field lines of opposite polarity draw together from either side of a closed magnetic field region.

expanding pressure of the superheated gas blows the field lines open and, with only one foot tied to the Sun, the unipolar lines are dragged into interplanetary space with the outrushing plasma. Thermal energy that once heated the trapped plasma now serves to accelerate it. By the time the escaping gas has traveled the equivalent of a few solar radii, it has reached an average velocity of 900,000 miles per hour—eight times the speed of sound in plasma.

As shown above, the speed and density of the solar wind vary with where and how it issues from the corona. At so-called coronal holes, the regions where normally looped magnetic lines open straight into space, the expanding plasma jets out in high-speed, low-density streams. Less forceful and somewhat denser are so-called helmet streamers (for their resemblance to World War I German helmets)—small streams of relatively quiet wind arising from magnetic lines that arch high above regions of closed magnetism, where magnetic lines stay fairly near the surface. Perhaps the most spectacular form of the wind is what is known as a coronal mass ejection, a huge, rapidly expanding dense cloud of plasma. Magnetic forces associated with shock waves at the front edge of the more violent of these eruptions can accelerate charged particles almost to the speed of light.

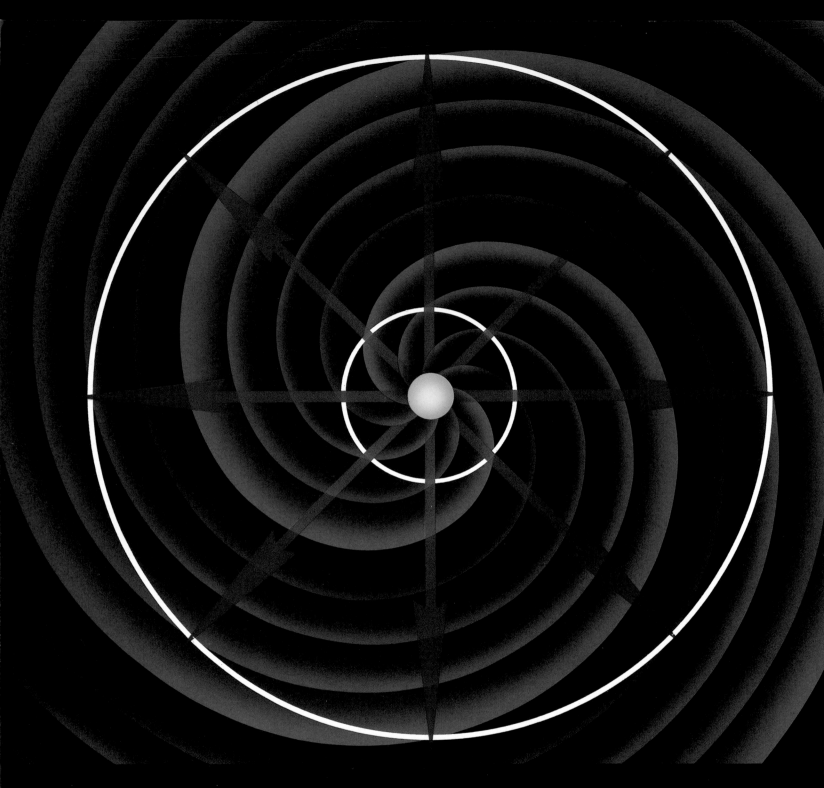

INTERPLANETARY LINES OF FORCE

As the coronal plasma rushes into interplanetary space, the open lines of magnetic force that act as its channel of escape are carried with it by the plasma's high electrical conductivity. Entrained in the solar wind, this magnetism constitutes the so-called interplanetary magnetic field. Although relatively weak— at Earth's orbit it is one ten-thousandth as strong as Earth's own magnetic field—the interplanetary field has a powerful influence on the frequency and severity of magnetic storms throughout the Solar System.

As shown above in a schematic overhead view, one end of each of the open field lines *(purple)* that make up the interplanetary field remains firmly embedded in the Sun; thus, the lines become twisted as the Sun rotates, creating a field that resembles the spiral in a nautilus shell. Meanwhile, the solar wind *(arrows)*, radiating from the Sun in all directions, traces a straight-line path relative to the spiral field, much as the needle of a phonograph travels radially across the grooves of a record. At the orbit of Earth *(inner white circle)*, the angle of the interplanetary field to the solar

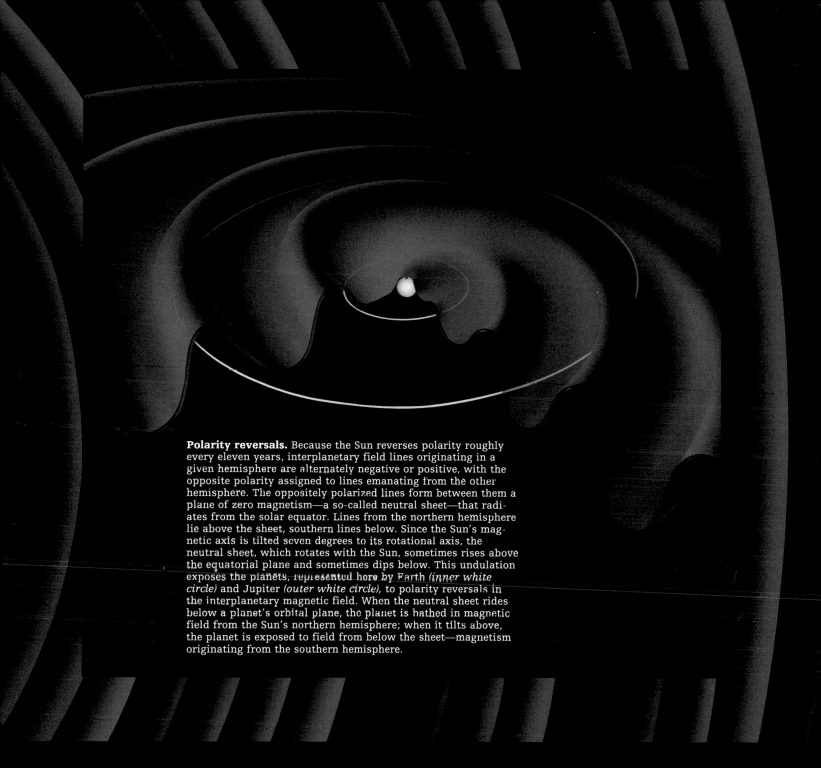

Polarity reversals. Because the Sun reverses polarity roughly every eleven years, interplanetary field lines originating in a given hemisphere are alternately negative or positive, with the opposite polarity assigned to lines emanating from the other hemisphere. The oppositely polarized lines form between them a plane of zero magnetism—a so-called neutral sheet—that radiates from the solar equator. Lines from the northern hemisphere lie above the sheet, southern lines below. Since the Sun's magnetic axis is tilted seven degrees to its rotational axis, the neutral sheet, which rotates with the Sun, sometimes rises above the equatorial plane and sometimes dips below. This undulation exposes the planets, represented here by Earth *(inner white circle)* and Jupiter *(outer white circle)*, to polarity reversals in the interplanetary magnetic field. When the neutral sheet rides below a planet's orbital plane, the planet is bathed in magnetic field from the Sun's northern hemisphere; when it tilts above, the planet is exposed to field from below the sheet—magnetism originating from the southern hemisphere.

wind averages 45 degrees; at Jupiter *(outer white circle)*, it is almost 90 degrees. Thus, at greater distances from the Sun, some of the more energetic charged particles are deflected by the coiled lines of force and follow the ever-tightening spiral into space.

The polarity, or magnetic direction, of the field—a key factor in the generation of magnetic disturbances—changes according to the relative orientation of field and planet. It can be either positive or negative, as defined by where a compass needle points when aligned with the field's source: When the needle points away from the Sun, the field is positive; when toward the Sun, it is negative. The field's polarity at any orbit

depends on whether the field lines sweeping past the planet originate in the Sun's northern or southern hemisphere *(box, above)*. When the direction of the interplanetary field is opposite to the planet's magnetic field, a process known as magnetic merging occurs, triggering aurorae and other electromagnetic disturbances. At Earth, the interplanetary field's polarity may reverse as many as four times during a twenty-seven-day solar rotation. Scientists acknowledge that this model is valid only during solar minimum; when sunspot activity is at its height, both the angular path and the polarity of the field become more chaotic and less amenable to descriptive modeling.

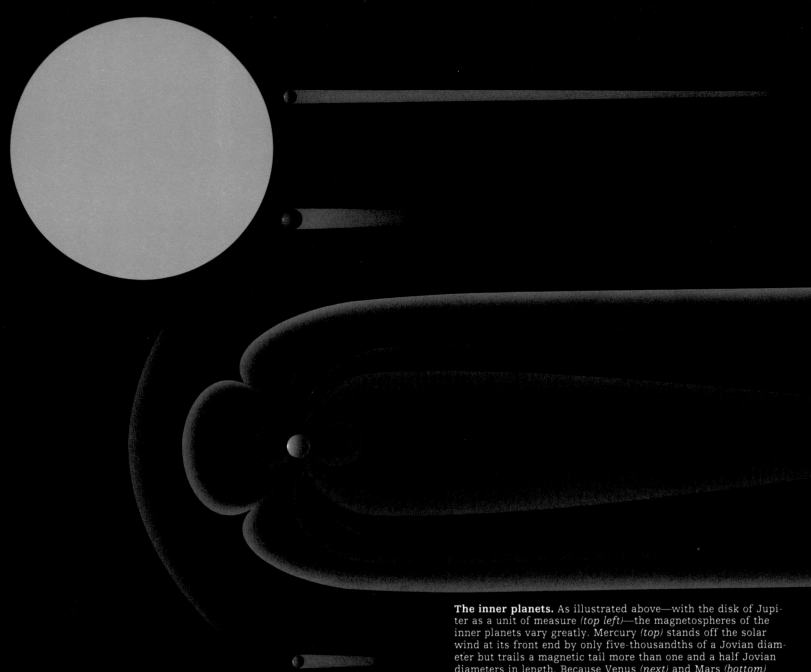

The inner planets. As illustrated above—with the disk of Jupiter as a unit of measure *(top left)*—the magnetospheres of the inner planets vary greatly. Mercury *(top)* stands off the solar wind at its front end by only five-thousandths of a Jovian diameter but trails a magnetic tail more than one and a half Jovian diameters in length. Because Venus *(next)* and Mars *(bottom)* have little or no intrinsic magnetism, their envelopes are mere fractions wider than the planets themselves. But Earth *(third from top)* is immersed in a magnetosphere as wide as Jupiter and fifty Jovian diameters long (eleven feet at this scale).

Magnetospheres: A Cosmic Standoff

Just as buildings force wind to flow around and over them, the planets interrupt the passage of the solar particles, creating windless hollows on their lee sides. The wind in turn sculpts these cavities into magnetospheres, great windsock shapes that grow and shrink according to the relative strengths of the planet's own magnetic field and the pressure exerted by the solar plasma, a function of its density at that planet's orbit.

A planet's magnetosphere acts as a kind of shield against the electrically charged particles of the wind. Its ability to repel the solar wind may be judged by the distance between the leading edge of the magnetosphere and the planet itself. For example, not only does Mercury have a weak magnetic field, but its atmosphere, which could also act as a buffer, is virtually nonexistent. Thus, the wind—at its most dense so close to the Sun and hence at its most powerful—can at times ram the magnetosphere almost to Mercury's surface. By contrast, Jupiter's magnetosphere is so charged with magnetic force that the solar wind, some 200 times less dense than at Mercury, rarely pushes

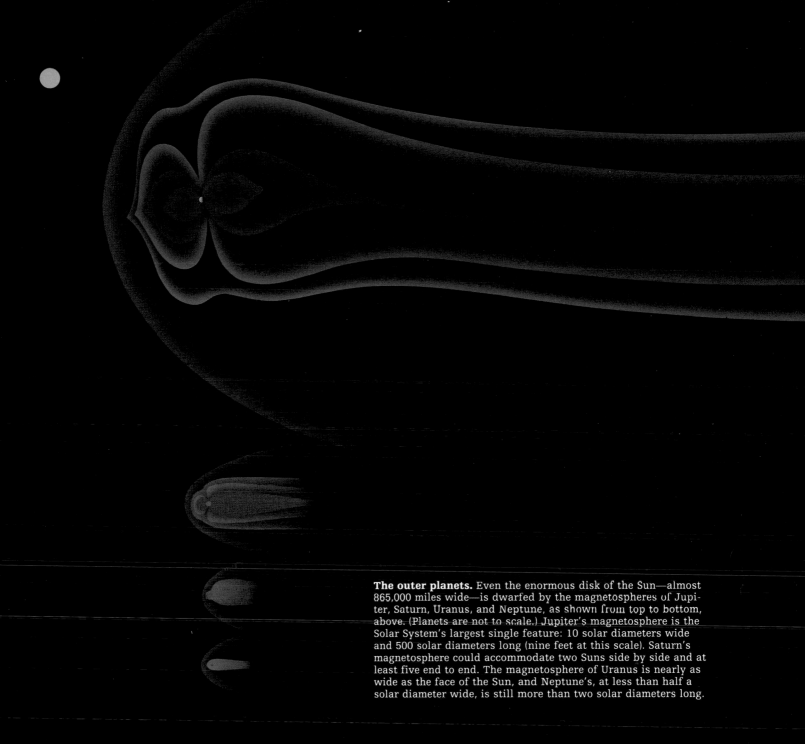

The outer planets. Even the enormous disk of the Sun—almost 865,000 miles wide—is dwarfed by the magnetospheres of Jupiter, Saturn, Uranus, and Neptune, as shown from top to bottom, above. (Planets are not to scale.) Jupiter's magnetosphere is the Solar System's largest single feature: 10 solar diameters wide and 500 solar diameters long (nine feet at this scale). Saturn's magnetosphere could accommodate two Suns side by side and at least five end to end. The magnetosphere of Uranus is nearly as wide as the face of the Sun, and Neptune's, at less than half a solar diameter wide, is still more than two solar diameters long.

the force shield closer than thirty-five Jovian diameters from the planet.

The role of the wind's density in changing the size of a planet's magnetosphere is clear at Saturn. Saturn's magnetic field is thirty-two times less powerful than Jupiter's, but the solar wind pressure at Saturn is only a quarter the pressure at Jupiter, allowing Saturn's weak magnetosphere to expand. Though still smaller than Jupiter's, it is vast enough to encompass several solar disks.

The most active magnetospheric contortions occur at Uranus and Neptune, both of whose magnetic axes are greatly displaced from their rotational axes—60

degrees and 50 degrees, respectively, compared with 10 degrees each for Earth and Jupiter. This variance, in turn, causes their magnetospheres to expand and dwindle continually as the magnetic axes vary in their degree of alignment with the solar wind.

Finally, even Mars and Venus, which have negligible or no magnetic fields, interact with the solar wind to create small magnetosphere-like cavities. The atmospheres of both planets are electrically conducting, creating enough resistance to cause the wind to slow down upon encountering them. Then the wind's plasma and interplanetary field lines are swept to the rear of the planets into a kind of tail.

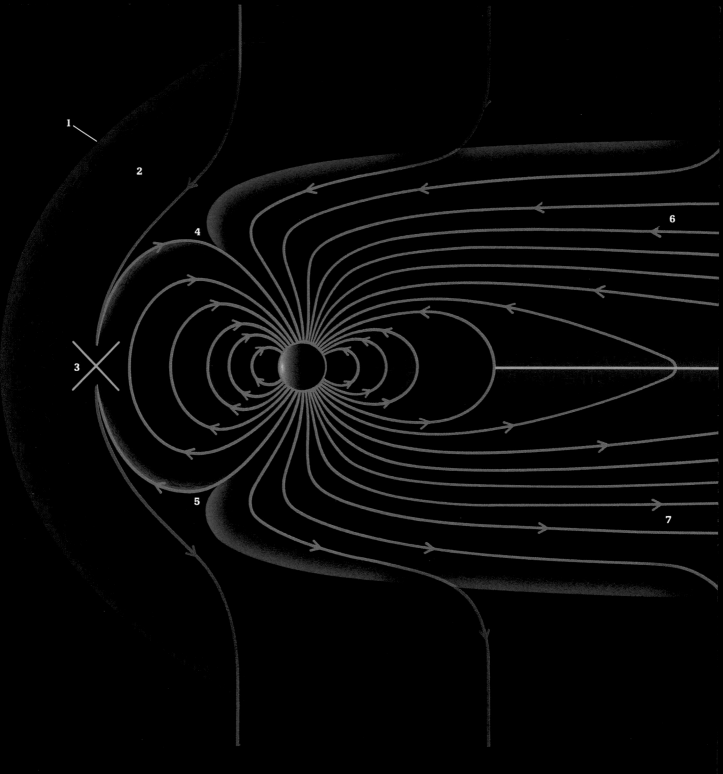

BREACHING EARTH'S LINES OF DEFENSE

The immense magnetic tails that form the bulk of a planet's magnetosphere are created and sustained by a complex process that enables the solar wind to break through the planet's magnetic defenses and transfer massive quantities of energy into the magnetosphere. At Earth *(above)*, for example, the solar plasma first encounters the bow shock *(1)*—a region rather like that around the nose of a supersonic aircraft, where complex interactions between gases and electromagnetic fields slow, compress, and heat the incoming wind, creating a zone of superheated charged particles called the magnetosheath *(2)*. When these particles hit the magnetopause *(3)*, the limit of Earth's magnetic influence, most of them are reflected back into space as if bouncing off an armored surface.

Periodically, however, when an undulation in the interplanetary field exposes Earth to magnetic field lines whose polarity is opposite to that of Earth's field, the two fields merge, creating a neutral point *(silver X)*

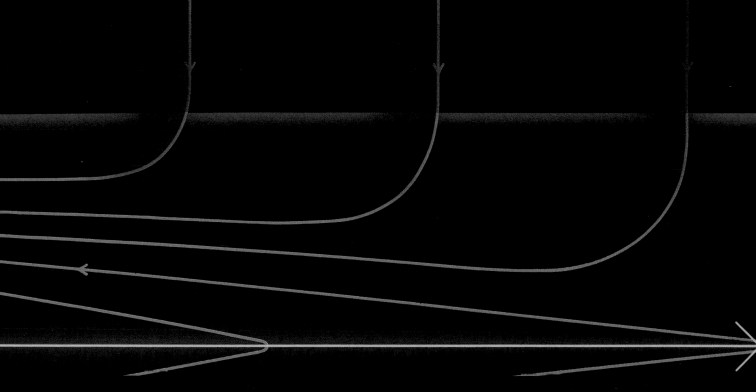

Genesis of the tail. In the first step of a sequence detailing the creation of the magnetotail, an interplanetary field line *(purple)* approaches a region *(silver X)* where it can merge magnetically with Earth's field *(blue).*

The interplanetary field line meets an Earth field line at the center of the X, a neutral point where the magnetic field strength is near zero. The magnetic field lines then begin to diffuse into one another.

The weakened field lines snap in half, releasing magnetic energy that propels them away from the neutral point. Each half of the interplanetary line magnetically merges with half of an oppositely polarized Earth field line, creating two hybrid lines.

With one end rooted in Earth's magnetic poles and the other in the solar plasma, the hybrid lines are swept back to become part of the magnetotail. The process begins anew as another interplanetary line nears the region where magnetic merging occurs.

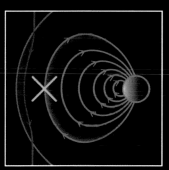
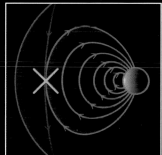
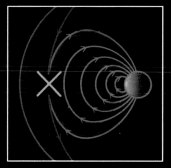
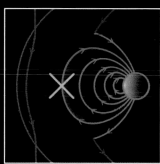

on the magnetopause *(3).* The merged, or hybrid, lines explode away from the neutral point, propelled by energy released during the merging process, and are dragged by the flow of the solar wind across both poles *(4, 5)* and into the region behind Earth to form the long magnetotail *(box, above).*

The magnetosphere is now effectively open to the solar wind: Each time a hybrid line is swept back into the tail, solar plasma flows along the field into Earth's magnetosphere, generating a current that in turn injects enormous amounts of electricity into the system. Electromagnetic forces drive the plasma-

laden field lines from the top and bottom halves of the magnetotail *(6, 7),* transporting the energy toward the tail's midplane. Where the oppositely polarized field lines touch, they create a new neutral point *(8),* releasing magnetic energy and propelling plasma earthward as well as into interplanetary space. Within minutes, the process deposits the equivalent of several days' U.S. energy consumption in the inner magnetosphere, where it is stored for an hour or more. Then, in a paroxysm of cosmic proportions, it is unleashed into the solar wind and into Earth's atmos-

phere *(pages 132-133).*

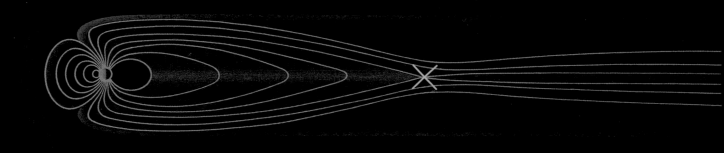

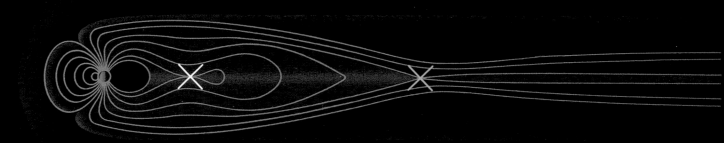

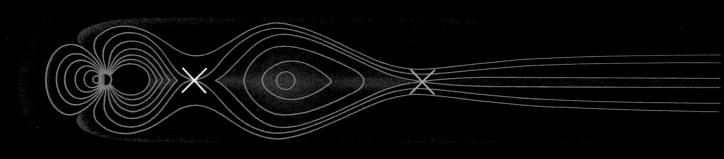

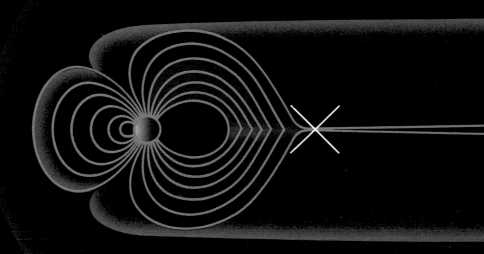

Spawn of the Plasma Wind

Except for radio interference and the eerie brilliance of the auroral displays that illuminate the polar night, the magnetic storms that regularly rage in Earth's magnetosphere are undetectable from the surface of the planet, 100,000 miles below. An average of four times a day, the entire magnetic envelope surrounding Earth contorts and releases billions of kilowatt-hours of stored energy. Such disturbances, called substorms, return the magnetosphere to a more stable, low-energy state.

During the main phase of the storm, stored magnetic energy is converted to kinetic energy through magnetic merging deep in the magnetotail *(diagrams, left)*.

This kinetic energy gives rise to one of the most spectacular and bizarre phenomena in the cosmos: the catapulting of an enormous egg of superheated plasma into space at speeds that can exceed two million miles per hour. The snap also hurtles electrons and ions from the magnetotail toward Earth along the magnetic field lines connected to the atmosphere at high latitudes. Bombarded by the electrons, molecules of atmospheric gases break up into atoms, emitting crimson, green, and purple arcs of light in the process. After about three hours, the storm subsides and the magnetosphere returns to its presubstorm configuration *(pages 130-131)*.

Buildup. About one hour before the onset of a substorm, field lines stretch and the magnetotail swells *(top left)* as the merging of interplanetary and Earth magnetic fields transfers more and more solar wind energy to the magnetosphere. The tail's midsection thins in response, and plasma-laden field lines migrate rapidly inward *(middle)*. The oppositely polarized sides of the first magnetic loop touch, break, and reconnect, creating a substorm neutral point *(white X)* upstream of the neutral point back in the tail *(silver X)*. More lines follow suit *(bottom)*, forming a set of closed field lines at Earth and a magnetic island, or plasmoid, downstream. From either side of the substorm neutral point come jets of highly energized particles; on one side, electrons pour into Earth's ionosphere to power the aurorae; on the other side, the plasmoid is enveloped in a sheath of magnetized, superhot plasma.

Expulsion. As the final set of tail lines meet at the substorm neutral point, stupendous amounts of energy are released. Like a magnetic slingshot, the merged lines hurl the giant bubble of plasma—some seventy-five Earth radii long, twenty Earth radii wide, and twelve Earth radii high—tailward into interplanetary space.

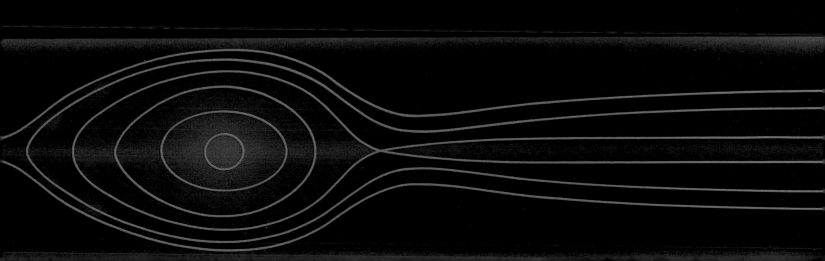

GLOSSARY

Absorption line: a dark line or band at a particular wavelength on a spectrum, formed when a substance between a radiating source and an observer absorbs electromagnetic radiation of that wavelength. Different substances produce characteristic patterns of absorption lines.

Angstrom: a unit of length equal to one ten-billionth of a meter, or about four-billionths of an inch; used in astronomy as a measure of wavelength.

Antineutrino: a subatomic particle, produced during fusion reactions, that transforms neutrons into protons. Its detection is used to confirm the existence of neutrinos. *See* Neutrino.

Arc second: one-sixtieth of an arc minute, which is in turn a sixtieth of a degree of arc; there are 360 degrees in a circle. In solar astronomy, arc seconds are used to measure the smallest discernible features of the Sun.

ATP (adenosine triphosphate): energy-storing molecules created by living organisms through the conversion of nutrients, such as sugar, and used to fuel all cells. In green plants, specialized biological structures called chloroplasts help convert sunlight into ATP.

Aurora: light given off by collisions between charged particles trapped in a planet's magnetic field and atoms of atmospheric gases near the planet's magnetic poles. Aurorae are visible on Earth as the aurora borealis, or northern lights, and the aurora australis, or southern lights.

Bolometer: an instrument that measures radiant energy, including microwave and infrared radiation.

Bow shock: the boundary region of interplanetary space where the solar wind is first deflected by a planet's magnetic field.

Brightness: in astronomy, the amount of light received on Earth from an object in space. The object's brightness is a combined result of its actual luminosity, its distance, and any light absorption by intervening dust or gas.

Butterfly diagram: a plot of sunspot latitude as a function of time, showing the latitudes at which sunspots were observed in the course of a century or more. *See* Sunspot cycle.

Calorie: a measurement denoting the amount of heat needed to raise the temperature of one gram of water by one degree Celsius at one atmospheric pressure, from a specified initial temperature.

Charge-coupled device (CCD): an electronic array of detectors, usually positioned at a telescope's focus, that registers electromagnetic radiation.

Charged particles: Fundamental components of matter, such as protons and electrons, that are responsible for all electrical phenomena. Charged particles are designated as positive or negative; each is surrounded by an electrical force field that attracts particles of opposite charge and repels those of like charge.

Chlorophyll: any of a group of related green pigments, found only in plants, that are essential for the process of photosynthesis. *See* Photosynthesis.

Chloroplast: An organelle, or specialized biological structure, that helps convert sunlight into energy during photosynthesis. Chloroplasts are found in all plant cells.

Chromosphere: the lower layer of the Sun's atmosphere, lying between the photosphere and the corona. Its name, meaning "sphere of color," derives from its appearance as a bright red ring encircling the lunar silhouette just before and after the totality of a solar eclipse.

Convection zone: the 125,000-mile-deep region just below the Sun's surface where energy is transported outward by moving streams of gas in a process called convection.

Core: in solar astronomy, the heart of the Sun, where energy is generated by nuclear reactions. *See* Fusion.

Corona: the outer layer of the Sun's atmosphere, composed of highly diffused, superheated, ionized gases.

Coronagraph: a solar telescope in which an occulting, or blocking, disk produces an artificial eclipse of the Sun, enabling the instrument to photograph the corona.

Coronal hole: any short-lived, low-density region of the Sun's corona through which the solar wind escapes.

Coronal mass ejection (CME): a vast bubble of plasma that erupts from the Sun's corona and travels through space at high speed.

Corpuscular radiation: a stream of electrically charged particles hypothesized in the 1950s to emanate from the Sun; later renamed the solar wind.

Deuterium: a form of hydrogen whose nucleus contains one neutron and one proton; also known as heavy hydrogen.

Deuteron: The nucleus of the deuterium atom.

Differential rotation: the phenomenon whereby regions of a celestial body rotate at differing speeds about the core. The Sun revolves faster at its equator than at its poles.

Doppler shift: a change in the wavelength and frequency of sound or electromagnetic radiation, caused by the motion of the emitter, the observer, or both.

Eclipse: the obscuration of light from a celestial body as it passes through the shadow of another body. In a solar eclipse, the Moon passes between the Sun and the Earth, blocking (or occulting) the Sun. *See* Occulting disk.

Electromagnetic radiation: waves, ranging in length, of electrical and magnetic energy that travel through space at the speed of light.

Electromagnetism: the force that attracts oppositely charged particles and repels similarly charged particles. Electromagnetism does not affect neutral particles such as neutrinos.

Emission line: a bright band at a particular wavelength on a spectrum, emitted directly by the source, and revealing by its wavelength a chemical constituent of that source.

Fibrils: sinuous, horizontal strands of gas that appear on the Sun's surface, indicating an active region nearby.

Fission: a nuclear reaction that releases energy when heavyweight atomic nuclei break down into lighter nuclei. Fission powers so-called atomic bombs. *See* Fusion.

Flare: an explosive release, marked by a sudden brightening near a sunspot or prominence, of electromagnetic radiation and huge quantities of charged particles from a small area of the solar surface.

Flux: the flow, or intensity, of matter or energy that either is a fluid or can be considered as a fluid.

Flux tubes: Magnetic tunnels, approximately 300 miles in diameter, that run horizontally beneath the surface of the Sun around the time of solar maximum. They form when magnetic field lines drawn close by differential rotation suddenly join together.

Frequency: the number of oscillations per second of an electromagnetic wave or other wave. *See* Wavelength.

Fusion: the combining of two atomic nuclei to form a heavier nucleus, with energy released as a by-product. Fusion powers so-called hydrogen bombs. *See* Fission.

Gamma rays: the most energetic form of electromagnetic

radiation, with the highest frequency and the shortest wavelength. Because Earth's atmosphere absorbs most of the radiation at that end of the spectrum, gamma ray studies of the Sun are usually conducted from space.

Gauss: the unit of measurement of a magnetic field. Earth's magnetic field, for example, averages less than one gauss, while the Sun's reaches almost 5,000 in isolated regions such as sunspots.

Geomagnetic storm: a series of terrestrial disturbances—namely, the precipitation of aurorae and rapid changes in the Earth's magnetic field—caused by high-speed blasts of the solar wind. *See* Substorm.

Grana: stacks of thin membranes, essential to the conversion of sunlight into energy, that are located inside plant-cell components called chloroplasts.

Granule: a roughly circular region on the Sun whose bright center indicates hot gases rising to the surface and whose dark edges indicate cooled gases that are descending toward the interior; granules give the photosphere its dappled look.

Helio: a prefix meaning "of or pertaining to the Sun," from the Greek word for the Sun, *helios*.

Helioseismology: the study of the Sun's surface oscillations as they relate to the internal structure and dynamics of the star.

Heliosphere: the vast volume of interplanetary space defined by the solar wind and its associated magnetic field, which extend to the edge of the Solar System and possibly beyond.

Heliostat: a solar-telescope component in which a mirror under computer control is automatically adjusted to follow the Sun in its path across the sky; this reflects the sunlight into the telescope at a constant angle.

Helmet streamer: a jet of low-speed solar-wind plasma —so named for its shape, which resembles the spike on a World War I German helmet—that flows outward from the Sun's corona.

Infrared: a band of electromagnetic radiation with a lower frequency and a longer wavelength than visible red light. Most of the Sun's infrared radiation is absorbed by Earth's atmosphere, but certain wavelengths can be detected from Earth.

Interplanetary magnetic field (IMF): an electromagnetic envelope sculpted around the Sun by the solar wind and believed to extend beyond the limits of the Solar System. *See* Magnetic field.

Ion: an atom that has lost or gained one or more electrons, thus becoming electrically charged. By contrast, a neutral atom has an equal number of negatively charged electrons and positively charged protons, giving the atom a zero net electrical charge.

Ionization: the process by which an electrically neutral atom loses or gains electrons.

Ionosphere: the area of Earth's atmosphere that extends from about 50 to 300 miles above the surface of the planet and is made up of multiple layers dominated by electrically charged, or ionized, atoms.

Isotope: one of two or more atomic variants of a chemical element, having the same number of protons as the most common form of the element but a differing number of neutrons.

Limb: the apparent edge of the Sun (or Moon) as it is seen in the sky. Astronomers refer to the left edge of the solar disk as the Sun's east limb and to the right edge as its west limb.

Luminosity: an object's total energy output, usually measured in ergs, calories, or watts; sometimes called *intrinsic* or *absolute brightness*.

Magnetic field: a field of force around the Sun and the planets, generated by electrical currents, in which a magnetic influence is felt by other currents. The Sun's magnetic field, like that of Earth, exhibits a north and south pole linked by lines of magnetic force.

Magnetopause: the well-defined boundary between a planet's magnetosheath and its magnetosphere.

Magnetosheath: the turbulent region of interplanetary space where solar-wind particles are slowed and deflected around a planet's magnetosphere, or magnetic envelope. The magnetosheath is sandwiched between the bow shock and the magnetopause.

Magnetosphere: a large, energetic envelope of magnetic field lines shaped by interactions between a planet's magnetic field and the solar wind; the flow of charged particles from the Sun.

Magnetotail: a billowing, streamlike extension of a planet's magnetosphere formed on the planet's dark side by the action of the solar wind.

Mass: a measure of the total amount of material in an object, determined by the object's gravity or by its tendency to resist acceleration.

Maunder minimum: the period from 1645 to 1715, when the Sun was virtually devoid of sunspots. The seventy-year stretch of minimum solar activity was discovered by British astronomer Edward Maunder in 1890.

Muon: a charged subatomic particle similar to an electron but much more massive and much less stable.

Neutrino: a chargeless, probably massless subatomic particle that is a by-product of thermonuclear reactions in the Sun's core. Neutrinos of various forms are sought by solar scientists for the details they may reveal about the processes fueling the Sun.

Objective lens: the principal lens of a telescope.

Occultation: the partial blocking of electromagnetic radiation from a celestial body by natural or mechanical means. *See* Coronagraph; Eclipse.

Occulting disk: a component of a solar telescope that blocks out the bright light from the disk of the Sun, making it possible for astronomers to study the comparatively weaker light from the Sun's corona.

Photon: a packet of electromagnetic energy that behaves like a chargeless particle traveling at the speed of light.

Photosphere: the visible, intensely bright, 200-mile-deep "surface" of the Sun. It is located between the convection zone and the chromosphere and differs from the chromosphere in that its constituent gases are stronger than, rather than subject to, their associated magnetic fields.

Photosynthesis: the process by which green plants use solar energy to manufacture carbohydrates and other organic nutrients from carbon dioxide and water. A critical by-product of photosynthesis is oxygen.

Photovoltaic (PV) energy: electrical energy produced by the exposure of a chemical compound—typically, a block of silicon, boron, and phosphorus—to sunlight.

Plage: from the French word for "beach," a bright, dense cloud of chromospheric gases found hovering above sunspots or other active areas of the solar surface.

Plasma: a gaslike conglomeration of charged particles that respond collectively to electrical currents and magnetic fields. Considered a fourth state of matter (along with solids, liquids, and gases), plasma constitutes the bulk of the Sun.

Plasmoid: a body of plasma pinched off the end of a planet's magnetotail by magnetic pressure from the solar wind.

Polarity: the intrinsic polar orientation, alignment, or separation of a force or physical property. The Sun displays an overall magnetic polarity, which reverses itself every eleven years or so.

Polarization: the state in which electromagnetic waves, usually beams of visible light, oscillate in only one, rather than in all, planes or directions.

Positron: a subatomic particle similar in mass to an electron but carrying a positive electric charge.

Prime focus: the spot inside a telescope where the electromagnetic radiation collected by the instrument is initially directed.

Prominence: an outcropping of solar gases suspended in the Sun's chromosphere or corona by magnetic forces. Active prominences resemble loops, eruptive prominences may arch two million miles outward before bursting, and quiescent prominences tend to linger in one place uneventfully for weeks or months.

Proton-proton reaction: a sequence of fusion reactions in which hydrogen nuclei, consisting of single protons, fuse to form deuterium and, ultimately, helium, releasing energy in the process; also called the proton-proton chain.

Pyrheliometer: a nineteenth-century instrument used to measure the total amount of radiant energy that the Sun delivers to a square centimeter of the Earth in one minute.

Quantum mechanics: a mathematical description of the rules by which subatomic particles interact, decay, and form atomic or nuclear objects. It is based on the quantum principle that energy is emitted in discrete, rather than continuous, units.

Radiative zone: an interior layer of the Sun, lying between the core and the convection zone, where energy travels outward by the process of radiation.

Radiometer: any device that measures the intensity of electromagnetic radiation.

Resolution: the degree to which details in an image can be separated, or resolved. The resolving power of a telescope is usually proportional to the diameter of its mirror or aperture. *Spatial resolution* refers to the ability to discriminate features, *spectral resolution* to the ability to monitor very narrow spectral bands.

Shock wave: a sudden discontinuity in the flow of a gas, a liquid, or a plasma, characterized by abrupt changes in the temperature, pressure, and velocity of the matter.

Solar activity cycle: the period of approximately twenty-two years between the appearance of sunspots having the same latitude and magnetic polarity. *See* Sunspot cycle.

Solar constant: the average amount of energy, usually measured in watts, that the Sun delivers to a square meter outside Earth's atmosphere at one astronomical unit (the Earth's mean distance from the Sun).

Solar maximum: the peak of a sunspot cycle, when the number of sunspots reaches its maximum before starting to subside.

Solar minimum: the beginning or end of a sunspot cycle, marked by the near absence of sunspots.

Solar wind: a continuous current of charged particles that streams outward from the Sun through the Solar System.

Spectral line: a bright or dark band in an astronomical spectrum that is produced by atoms as they absorb or emit light.

Spectrograph: an instrument that splits light or other electromagnetic radiation into its individual wavelengths, collectively known as a spectrum, and records the result photographically or electronically. When the instrument lacks such a recording capability, it is called a spectroscope.

Spectroheliogram: a photograph of the Sun captured by a spectroheliograph.

Spectroheliograph: an instrument used to photograph the Sun in the monochromatic light of a selected wavelength; in essence, a special-purpose spectrograph.

Spectrometer: a spectroscope that has been fitted with scales to measure the positions of various spectral lines.

Spectroscope: any of various instruments used for the direct observation of a spectrum.

Spectroscopy: the study of spectra, including the position and intensity of emission and absorption lines, to determine the chemical elements or physical processes that created them.

Spectrum: the array of electromagnetic radiation, arranged in order of wavelength from long-wave radio emissions to short-wave gamma rays; also, a narrower band of wavelengths, called the visible spectrum, as when light dispersed by a prism shows its component colors. Spectra are often striped with emission or absorption lines, which can be examined to reveal the composition and motion of the light source.

Spicule: a short-lived, narrow, vertical jet of gas that originates on the solar surface and rises as high as 6,000 miles into the chromosphere.

Substorm: a magnetic disturbance caused when Earth's magnetosphere expels energy that has built up inside it from the solar wind. Occurring about four times a day, substorms precipitate aurorae—the only visible evidence of a substorm's passing. *See* Geomagnetic storm.

Sunspot: a dark, fringed blemish on the solar surface that is caused by a concentration of the Sun's magnetic field lines.

Sunspot cycle: the recurring, eleven-year rise and fall in the number of sunspots.

Totality: the state of total solar eclipse, when the entire sphere of the Sun is obscured by the Moon.

Ultraviolet (UV): a band of electromagnetic radiation with a higher frequency and a shorter wavelength than visible blue light. Because Earth's atmosphere absorbs most ultraviolet emissions, thorough studies of the Sun's UV output must be conducted from space.

Watt: a unit of electrical power equal to 0.0013406 horsepower.

Wavelength: the distance from crest to crest or trough to trough of an electromagnetic or other wave. Wavelengths are related to frequency: The longer the wavelength, the lower the frequency.

X-ray: a band of electromagnetic radiation intermediate in wavelength between ultraviolet and gamma rays. Because x-rays are absorbed by Earth's atmosphere, x-ray astronomy is performed in space.

Zeeman effect: a broadening or splitting of spectral lines caused by the influence on the light source of a powerful magnetic field.

BIBLIOGRAPHY

Books

Abell, George O., *Drama of the Universe*. New York: Holt, Rinehart and Winston, 1978.

Abell, George O., David Morrison, and Sidney C. Wolff, *Exploration of the Universe*. Philadelphia: Saunders College Publishing, 1987.

Akasofu, S. I., and Y. Kamide, eds., *The Solar Wind and the Earth*. Dordrecht: D. Reidel, 1987.

Alter, Dinsmore, Clarence H. Cleminshaw, and John G. Phillips, *Pictorial Astronomy*. New York: Harper & Row, 1983.

Asimov, Isaac:
Asimov's Guide to Science, Vol. 1: The Physical Sciences. New York: Penguin Books, 1982.
The Sun Shines Bright. New York: Avon Books, 1981.

Beatty, J. Kelly, Brian O'Leary, and Andrew Chaikin, eds., *The New Solar System*. Cambridge, Mass.: Sky, 1982.

Berman, Louis, and J. C. Evans, *Exploring the Cosmos*. Boston: Little, Brown, 1977.

Brandt, John C.:
Comets. San Francisco: W. H. Freeman, 1981.
Introduction to the Solar Wind. San Francisco: W. H. Freeman, 1970.

Carovillano, R. L., and J. M. Forbes, eds., *Solar-Terrestrial Physics*. Dordrecht: D. Reidel, 1983.

Chaisson, Eric, *Universe: An Evolutionary Approach to Astronomy*. Englewood Cliffs, N.J.: Prentice-Hall, 1988.

Close, Frank, Michael Marten, and Christine Sutton, *The Particle Explosion*, New York: Oxford University Press, 1987.

Dixon, Robert T., *Dynamic Astronomy*. Englewood Cliffs, N.J.: Prentice-Hall, 1989.

Eather, Robert H., *Majestic Lights: The Aurora in Science, History, and the Arts*. Washington, D.C.: American Geophysical Union, 1980.

Eddy, John A., *The New Sun: The Solar Results from Skylab*. Washington, D.C.: NASA, 1979.

Eddy, John A., ed., *The New Solar Physics*. Boulder, Colo.: Westview Press, 1978.

Egeland, A., O. Holter, and A. Omholt, eds., *Cosmical Geophysics*. Oslo: Universitetsforlaget, 1973.

Field, George B., and Eric J. Chaisson, *The Invisible Universe: Probing the Frontiers of Astrophysics*. Boston: Birkhauser, 1985.

Fire of Life. New York: Smithsonian Exposition Books, 1981 (distributed by W. W. Norton, New York).

Frazier, Kendrick, *Our Turbulent Sun*. Englewood Cliffs, N.J.: Prentice-Hall, 1980.

Friedman, Herbert, *Sun and Earth*. New York: Scientific American Books, 1986.

Giovanelli, Ronald G., *Secrets of the Sun*. Cambridge: Cambridge University Press, 1984.

Hellman, Geoffrey T., *The Smithsonian: Octopus on the Mall*. Philadelphia: J. B. Lippincott, 1967.

Herman, John R., and Richard A. Goldberg, *Sun, Weather, and Climate*. Washington, D.C.: NASA, 1978.

Herrmann, Dieter B., *The History of Astronomy from Herschel to Hertzsprung*. Revised and translated by Kevin Krisciunas. Cambridge: Cambridge University Press, 1984.

Hundhausen, A. J., *Coronal Expansion and Solar Wind*. New York: Springer-Verlag, 1972.

Jaber, William, *Exploring the Sun*. New York: Jul-

ian Messner, 1980.

Jager, C. de, and Z. Švestka, eds., *Progress in Solar Physics*. Dordrecht: D. Reidel, 1986.

Karttunen, H., et al., eds., *Fundamental Astronomy*. New York: Springer-Verlag, 1987.

Kennel, Charles F., Louis J. Lanzerotti, and E. N. Parker, eds., *Solar System Plasma Physics, Vol. 2: Magnetospheres*. Amsterdam: North-Holland, 1979.

Krane, Kenneth S., *Modern Physics*. New York: John Wiley & Sons, 1983.

Lang, Kenneth R., and Owen Gingerich, eds., *A Source Book in Astronomy and Astrophysics, 1900-1975*. Cambridge, Mass.: Harvard University Press, 1979.

Lewis, Richard S., *The Illustrated Encyclopedia of the Universe*. New York: Harmony Books, 1983.

Lockyer, Norman, *The Sun's Place in Nature*. London: Macmillan, 1897.

Meadows, A. J.:
Early Solar Physics. Oxford: Pergamon Press, 1970.
Science and Controversy: A Biography of Sir Norman Lockyer. Cambridge, Mass.: MIT Press, 1972.

A Meeting with the Universe. Washington, D.C.: NASA, 1981.

Merrill, Ronald T., and Michael W. McElhinny, *The Earth's Magnetic Field: Its History, Origin and Planetary Perspective*. London: Academic Press, 1983.

Mitton, Simon, *Daytime Star: The Story of Our Sun*. New York: Charles Scribner's Sons, 1981.

Moore, Patrick, *The Sun*. New York: W. W. Norton, 1968.

Nicolson, Iain, *The Sun*. New York: Rand McNally, 1982.

Nishida, A., *Geomagnetic Diagnosis of the Magnetosphere*. New York: Springer-Verlag, 1978.

Noyes, Robert W., *The Sun, Our Star*. Cambridge, Mass.: Harvard University Press, 1982.

Parker, E. N., *Interplanetary Dynamical Processes*. New York: John Wiley & Sons, 1963.

Parker, Sybil P., ed., *McGraw-Hill Encyclopedia of Astronomy*. New York: McGraw-Hill, 1983.

Pasachoff, Jay M., and Marc L. Kutner, *University Astronomy*. Philadelphia: W. B. Saunders, 1978.

Priest, E. R., *Solar Magneto-Hydrodynamics*. Dordrecht: D. Reidel, 1982.

Priest, E. R., ed., *Dynamics and Structure of Quiescent Solar Prominences*. Dordrecht: Kluwer Academic Publishers, 1989.

Protheroe, W., R. Cepriotti, and G. H. Newsom, *Exploring the Universe*. Columbus, Ohio: Charles E. Merrill, 1984.

Ridpath, Ian, *Stars and Planets*. London: Hamlyn, 1978.

Shapiro, Stuart L., and Saul A. Teukolsky, eds., *Highlights of Modern Astrophysics: Concepts and Controversies*. New York: John Wiley & Sons, 1986.

Shu, Frank H., *The Physical Universe: An Introduction to Astronomy*. Mill Valley, Calif.: University Science Books, 1982.

Stryer, Lubert, *Biochemistry*. San Francisco: W. H. Freeman, 1981.

Sturrock, Peter A., ed., *Physics of the Sun, Vol. 1: The Solar Interior*. Dordrecht: D. Reidel, 1986.

Tascione, Thomas F., *Introduction to the Space Environment*. Malabar, Fla.: Orbit, 1988.

Trefil, James, *Meditations at Sunset: A Scientist Looks at the Sky*. New York: Charles Scribner's Sons, 1987.

Washburn, Mark, *In the Light of the Sun*. New York:

Harcourt Brace Jovanovich, 1981.

Wright, Helen, *Explorer of the Universe: A Biography of George Ellery Hale.* New York: E. P. Dutton, 1966.

Zeilik, Michael, and Elske v. P. Smith, *Introductory Astronomy and Astrophysics.* Philadelphia: Saunders College Publishing, 1987.

Zirin, Harold:
Astrophysics of the Sun. Cambridge: Cambridge University Press, 1988.
The Solar Atmosphere. Waltham, Mass.: Blaisdell, 1966.

Zirker, Jack B., ed., *Coronal Holes and High Speed Wind Streams.* Boulder, Colo.: Colorado Associated University Press, 1977.

Periodicals

Akasofu, Syun-Ichi:
"The Aurora." *Scientific American,* December 1965.
"The Dynamic Aurora." *Scientific American,* May 1989.

"Another Successful OSO." *Sky & Telescope,* October 1969.

Arnon, Daniel I., "The Role of Light in Photosynthesis." *Scientific American,* November 1960.

Bagnall, Philip M., "Observe the Naked-Eye Sky Glows." *Astronomy,* June 1988.

Bahcall, J. N.:
"Neutrinos and Sunspots." *Nature,* November 26, 1987.
"Solar Neutrino Experiments." *Reviews of Modern Physics,* October 1978.

Barrett, Alan H., and Edward Lilley, "Mariner-2 Microwave Observations of Venus." *Sky & Telescope,* April 1963.

Bassham, J. A., "The Path of Carbon in Photosynthesis." *Scientific American,* June 1962.

Bernstein, Jeremy, "Out of My Mind." *American Scholar,* Spring 1988.

Chaikin, Andrew, "Solar Max: Back from the Edge." *Sky and Telescope,* June 1984.

Chapman, Robert W., "NASA's Search for the Solar Connection—1." *Sky & Telescope,* August 1979.

Christensen-Dalsgaard, Jorgen, and Douglas O. Gough, "Is the Sun Helium-Deficient?" *Nature,* December 11, 1980.

Christensen-Dalsgaard, Jorgen, Douglas O. Gough, and Juri Toomre, "Seismology of the Sun." *Science,* September 6, 1985.

De la Zerda Lerner, A., and K. O'Brien, "Atmospheric Radioactivity and Variations in the Solar Neutrino Flux." *Nature,* November 26, 1987.

Deming, Drake, et al., "Infrared Helioseismology: Detection of the Chromospheric Mode." *Nature,* July 17, 1986.

De Rújula, A., and S. L. Glashow, "Neutrino Weight Watching." *Nature,* August 21, 1980.

Dunn, Richard B., "High Resolution Solar Telescopes." *Solar Physics 100,* 1985.

"The Earth's Magnetic Tail." *Sky & Telescope,* June 1965.

Eberhart, Jonathan:
"Astronomers Plan a Month on the Sun." *Science News,* August 27, 1988.
"Solar Blast." *Science News,* May 27, 1989.

Eddy, John A., "The Case of the Missing Sunspot." *Scientific American,* May 1977.

Emslie, A. Gordon, "Explosions in the Solar Atmosphere." *Astronomy,* November 1987.

Evans, J. V., "The Sun's Influence on the Earth's Atmosphere and Interplanetary Space." *Science,* April 30, 1982.

"Fantastic Fortnight of Active Region 5395." *Science News,* April 8, 1989.

"The First Five Years of the Space Age." *Sky & Telescope,* March 1963.

Fisher, Arthur, "Hunting Neutrinos." *Popular Science,* May 1988.

Foukal, Peter, "Magnetic Loops in the Sun's Atmosphere." *Sky & Telescope,* December 1981.

Franco, Anna, and David H. Smith, "Vanishing Solar Neutrinos." *Sky & Telescope,* February 1987.

Giampapa, Mark S., "The Solar-Stellar Connection." *Sky & Telescope,* August 1987.

Golub, Leon, "What Heats the Solar Corona?" *Astronomy,* September 1982.

"Great Balls of Fire." *Sky & Telescope,* May 1989.

Grec, Gérard, Eric Fossat, and Martin Pomerantz, "Solar Oscillations: Full Disk Observations from the Geographic South Pole." *Nature,* December 11, 1980.

Gribbin, John, "Gravity, Dust and Solar Neutrinos." *Astronomy,* June 1978.

Gringauz, K. I., "Some Results of Experiments in Interplanetary Space by Means of Charged Particle Traps on Soviet Space Probes." *Space Research,* 1961, vol. 2, pp. 539-553.

Harvey, J., J. R. Kennedy, and J. W. Leibacher, *"GONG: To See Inside Our Sun." Sky & Telescope,* November 1987.

Harvey, J., M. Pomerantz, and T. Duvall, Jr., "Astronomy on Ice." *Sky & Telescope,* December 1982.

Harvey, J., et al., "The Global Oscillation Network Group *(GONG)." Advances in Space Research,* 1988, vol. 8, no. 11, pp. 117-120.

Hones, Edward W., Jr., "The Earth's Magnetotail." *Scientific American,* March 1986.

"IMP Discovers New Radiation Belt." *Missiles and Rockets,* March 23, 1964.

Jaroff, Leon, "Fury on the Sun." *Time,* July 3, 1989.

Kanipe, Jeff, Richard Talcott, and Robert Burnham, "The Rise and Fall of the Sun's Activity." *Astronomy,* October 1988.

Kidder, Tracy, "A Blemished Sun?" *Science,* July-August 1981.

Klein, L. W., and L. F. Burlaga, "Interplanetary Magnetic Clouds at 1 AU." *Journal of Geophysical Research,* February 1, 1982.

Kopp, Roger A., "The High Altitude Observatory's 1970 Eclipse Expedition." *Sky & Telescope,* June 1970.

Krieger, A. S., A. F. Timothy, and E. C. Roelof, "A Coronal Hole and Its Identification as the Source of a High Velocity Solar Wind Stream." *Solar Physics,* 1973, vol. 29, pp. 505-525.

Kuhn, J. R., K. G. Libbrecht, and R. H. Dicke, "The Surface Temperature of the Sun and Changes in the Solar Constant." *Science,* November 1988.

Lanzerotti, Louis J., and Stamatios M. Krimigis, "Comparative Magnetospheres." *Physics Today,* November 1985.

Leer, Egil, Thomas E. Holzer, and Tor Fla, "Acceleration of the Solar Wind." *Space Science Reviews,* 1982, vol. 33, no. 2, pp. 161-200.

Leibacher, John W., et al., "Helioseismology." *Scientific*

American, September 1985.

Libbrecht, K. G., et al., "The Excitation and Damping of Solar Oscillations." *Nature,* September 18, 1986.

Livingston, William, Oddbjorn Engvold, and Eberhart Jensen, "Old and New Views of Solar Prominences." *Astronomy,* July 1987.

LoPresto, James Charles:
"Looking Inside the Sun." *Astronomy,* March 1989.
"The Rotation of the Sun." *Astronomy,* December 1986.

McIntosh, Patrick S., and Harold Leinbach, "Watching the Premier Star." *Sky & Telescope,* November 1988.

MacRobert, Alan, "The Aurora." *Sky & Telescope,* January 1986.

Maran, Stephen P.:
"Cosmic Rays Meet the Solar Wind." *Natural History,* August 1983.
"Solar Bubbles." *Natural History,* April 1975.

Maran, Stephen P., and Bruce E. Woodgate, "A Second Chance for Solar Max." *Sky and Telescope,* June 1984.

Maxwell, Alan, "Solar Flares and Shock Waves." *Sky & Telescope,* October 1983.

Menzel, Donald H., and Jay M. Pasachoff, "Solar Eclipse, Nature's Super Spectacular." *National Geographic,* August 1970.

Morrison, Philip, "The Neutrino." *Scientific American,* January 1956.

Mullan, Dermott:
"Caution! High Winds Beyond This Point." *Astronomy,* January 1982.
"Tuning in to the Interior of a Star." *Astronomy,* December 1984.

Nadis, Steve, "Stars." *Omni,* April 1989.

Ness, Norman F., Clcll S. Scearce, and Joseph B. Seek, "Initial Results of the Imp 1 Magnetic Field Experiment." *Journal of Geophysical Research,* September 1, 1964.

Neugebauer, Marcia, and Conway W. Snyder, "Mariner 2 Observations of the Solar Wind, 1, Average Properties." *Journal of Geophysical Research,* October 1, 1966.

"NOAO Scientist Reports on Signs of Violent 'Healing' Process on Sun." *National Optical Astronomy Observatories News,* May 13, 1985.

Noëns, J. C., and J. Pageault, "Measuring Electron Density in Coronal Active Regions." *Solar Physics,* 1984, vol. 94, pp. 117-131.

Norman, Eric B., "Neutrino Astronomy: A New Window on the Universe." *Sky & Telescope,* August 1985.

"Observing the Satellites." *Sky & Telescope,* May 1961.

Overbye, Dennis:
"The Ghost Universe of Neutrinos." *Sky & Telescope,* August 1980.
"John Eddy: The Solar Detective." *Discover,* August 1982.

Parker, E. N., "Dynamics of the Interplanetary Gas and Magnetic Fields." *Astrophysical Journal,* November 1958.

Peterson, Ivars:
"A Swirl on the Sun's Blotchy Face." *Science News,* September 17, 1988.
"New Pictures of the Sun Reveal a Number of Surprising and Puzzling Solar Features." *Science News,* July 1988.

"Probe Finds Unexpected Strength in Interplanetary Magnetic Field." *Aviation Week and Space Technology,* April 3, 1961.

Reeves, Hubert, "The Detection of Solar Neutrinos." *Sky & Telescope,* May 1964.

"Results from Mariner 2." *Sky & Telescope,* February 1963.

Roberts, B., and W. R. Campbell, "Magnetic Field Corrections to Solar Oscillation Frequencies." *Nature,* October 1986.

Robinson, Leif J., "The Sunspot Cycle: Tip of the Iceberg." *Sky & Telescope,* June 1987.

Roble, Raymond G., "The Auroras." *Natural History,* October 1977.

Russell, C. T., and R. L. McPherron, "The Magnetotail and Substorms." *Space Science Reviews,* 1973, vol. 15, pp. 205-264.

Rust, David M., "Warming Up for the Solar Maximum Year." *Sky & Telescope,* October 1979.

Saunders, Mark, "The Birth of a Plasmoid." *Nature,* June 29, 1989.

Severny, A. B., V. A. Kotov, and T. T. Tsap, "Observations of Solar Pulsations." *Nature,* January 15, 1976.

"Solar Max Snaps a Big, Brilliant Flare." *Science News,* March 18, 1989.

Stahl, Philip A., "Sunspots." *Astronomy,* May 1980.

Stevens, William K., "Analysis Links Sunspots to Weather on Earth." *New York Times,* June 13, 1989.

Thomsen, Dietrick E.:
"Neutrino Mass: A Positive View." *Science News,* April 18, 1987.
"Solar Neutrino Mysteries Persist." *Science News,* April 30, 1988.

Verschuur, Gerrit L.:
"The Many Faces of the Sun." *Astronomy,* March 1989.
"Will *Solar Max* Be Saved?" *Astronomy,* October 1988.

Waldmeier, M., "The Coronal Hole at the 7 March 1970 Solar Eclipse." *Solar Physics,* 1975, vol. 40, pp. 351-358.

Wang, Y. M., A. G. Nash, and N. R. Sheeley, Jr., "Magnetic Flux Transport on the Sun." *Science,* August 18, 1989.

Weisburd, S., "Mining for Traces of Galactic Star Deaths." *Science News,* June 4, 1988.

Weneser, Joseph, and Gerhart Friedlander, "Solar Neutrinos: Questions and Hypotheses." *Science,* February 13, 1987.

Wilcox, John M., and Norman F. Ness, "Structure." *Solar Physics,* 1967, vol. 1, pp. 437-445.

Willson, Richard C., and H. S. Hudson, "Solar Luminosity Variations in Solar Cycle 21." *Nature,* April 28, 1988.

Willson, Richard C., H. S. Hudson, and M. Woodard, "The Inconstant Solar Constant." *Sky & Telescope,* June 1984.

Other Sources

"Basic Photovoltaic Principles and Methods." Golden, Colo.: Solar Energy Research Institute, February 1982.

Harrington, Sherwood, "Selecting Your First Telescope." Information packet on astronomy. San Francisco: Astronomical Society of the Pacific, 1982.

Menicucci, David F., and Anne V. Poore, "Today's Photovoltaic Systems: An Evaluation of Their Performance." Albuquerque: Sandia National Laboratories, 1987.

Sofia, S., ed., "Variations of the Solar Constant." NASA Conference Publication 2191. Washington, D.C.: NASA Scientific and Technical Information Branch, 1981.

INDEX

ACKNOWLEDGMENTS

The editors wish to thank Ronald J. Angione, San Diego State University; Laurie Batchelor, NASA Goddard Space Flight Center, Greenbelt, Md.; Gerd Bauerfeld, Max Planck Institut für Astrophysik, Garching, West Germany; Ilse Biermann, Munich, West Germany; Ron Brashear, Mount Wilson Observatory, Pasadena, Calif.; Todd Brown, National Solar Observatory, Sunspot, N.Mex.; Alain Bücher, Observatoire du Pic-du-Midi, Observatoire Midi-Pyrénées, France; Leonard F. Burlaga, NASA Goddard Space Flight Center, Greenbelt, Md.; Canary Astrophysical Institute, Tenerife, Canary Islands, Spain; Kumar Chitre, NASA Goddard Space Flight Center, Greenbelt, Md.; Bill Cox, Smithsonian Institution Archives, Washington, D.C.; Steve Curtis, NASA Goddard Space Flight Center, Greenbelt, Md.; Raymond Davis, Jr.,

University of Pennsylvania, Philadelphia; John Eddy, University Corporation for Atmospheric Research, Boulder, Colo.; Robert Futaully, Observatoire du Pic-du-Midi, Observatoire Midi-Pyrénées, France; Sanjoy Ghosh, NASA Goddard Space Flight Center, Greenbelt, Md.; Peter Glaser, Arthur D. Little, Inc., Cambridge, Mass.; Leon Golub, Smithsonian Astrophysical Observatory, Cambridge, Mass.; Joe Gurman, NASA Goddard Space Flight Center, Greenbelt, Md.; Peter D. Hingley, Royal Astronomical Society, London; Art Hundhausen, National Center for Atmospheric Research, Boulder, Colo.; Fred Ipavich, University of Maryland, College Park; Larry Kazmerski, Solar Energy Research Institute, Golden, Colo.; Steve Keil, National Solar Observatory, Sunspot, N.Mex.; Ursula Karus, Hamburg, West Germany; Eugene Lavely, Massachusetts Institute of Technology, Cambridge; Kenneth Libbrecht, California Institute of Technology, Pasadena; William Livingston, National Solar Observatory, Tucson, Ariz.; Patrick McIntosh, National Oceanic and Atmospheric Administration, Boulder, Colo.; Donna McKinney, Naval Research Laboratory, Washington, D.C.; Richard Muller, Observatoire du Pic-du-Midi, Observatoire Midi-Pyrénées, France; Jacques-Clair Noëns, Observatoire du Pic-du-Midi, Observatoire Midi-Pyrénées, France; Lawrence J. November, National Solar Observatory, Sunspot, N.Mex.; William O'Clock, National Oceanic and Atmospheric Administration, Boulder, Colo.; Robert Pfaff, NASA Goddard Space Flight Center, Greenbelt, Md.; Theresa Salazar, University of Arizona, Tucson; Barbara Scott, NASA Goddard Space Flight Center, Greenbelt, Md.; Richard Sicher, Agricultural Research Service, United States Department of Agriculture, Beltsville, Md.; James Slavin, NASA Goddard Space Flight Center, Greenbelt, Md.; Marie-José Vin, Observatoire de Haute Provence, France; Richard Willson, NASA Jet Propulsion Laboratory, Pasadena, Calif.; Ed Zieske, Homestake Mining Co., Lead, S.Dak.; Harold Zirin, California Institute of Technology, Pasadena.

PICTURE CREDITS

The sources for the illustrations that appear in this book are listed below. Credits from left to right are separated by semicolons, from top to bottom by dashes.
Cover: Big Bear Solar Observatory, California Institute of Technology. 2, 3: Art by Rob Wood of Stansbury, Ronsaville, Wood, Inc. 8, 9: National Solar Observatory, Sunspot, N.Mex. 10, 11: Initial cap, detail from pages 8, 9—art by Matt McMullen. 13: Art by Fred Holz. 14: National Solar Observatory, Sunspot, N.Mex. 17: Lick Observatory, University of California, Santa Cruz. 18: From *The Life and Work of Sir Norman Lockyer,* by T. Mary Lockyer, Macmillan and Co., 1928, London—from *The Spectroscope and Its Applications,* by Norman Lockyer, Macmillan and Co., 1873, London. 20: The Observatories of the Carnegie Institution of Washington. 21: Yerkes Observatory, University of Chicago. 23: National Solar Observatory, Sunspot, N.Mex. Pullout artwork by Time-Life Books (2). 25: From *A New Sun,* by John Eddy, NASA, 1979, Washington, D.C. 27: National Solar Observatory, Sunspot, N.Mex. 29: Art by Stephen R. Bauer. Background photo, Big Bear Solar Observatory, California Institute of Technology. 30-39: Art by Stephen R. Bauer. 40, 41: Raymond Davis/Brookhaven National Laboratory. 42: Initial cap, detail from pages 40, 41. 44, 45: Background art by Stephen R. Wagner. Royal Astronomical Society, London; AIP Niels Bohr Library, Goudsmit Collection; AIP Niels Bohr Library, Margrethe Bohr Collection; courtesy the Archives, California Institute of Technology; AIP Niels Bohr Library, photo by Edith Michaelis; AIP Niels Bohr Library. 46: George B. Keeley/New Haven Register. 48: Observatoire du Pic-du-Midi. 50: David Parker, SPL, Photo Researchers. 54: Ken Libbrecht, California Institute of Technology—artwork by Time-Life Books. 55: National Optical Astronomy Observatories; Ken Libbrecht, California Institute of Technology. 56: Map by Time-Life Books. 58, 59: Background art by Stephen R. Wagner. National Optical Astronomy Observatories; Observatoire du Pic-du-Midi—National Optical Astronomy Observatories; Big Bear Solar Observatory, California Institute of Technology—National Optical Astronomy Observatories. 60, 61: Art by Stephen R. Wagner. National Optical Astronomy Observatories; inset art by Matt McMullen; National Optical Astronomy Observatories. 62, 63: Art by Stephen R. Wagner. National Solar Observatory, Sunspot, N.Mex. (3). 64, 65: Art by Stephen R. Wagner, based on drawings by Daniel Bardin. Observatoire du Pic-du-Midi (2). 66, 67: Art by Stephen R. Wagner. Big Bear Solar Observatory, California Institute of Technology (4). 68, 69: Art by Stephen R. Wagner. National Optical Astronomy Observatories (3). 70, 71: The Naval Research Laboratory. 72: Initial cap, detail from pages 70, 71. 74: Smithsonian Institution, no. A-10600; Smithsonian Institution, no. 80-4951. 76, 77: Courtesy Special Collections, University of Arizona; courtesy Royal Greenwich Observatory—James Balog/© 1982 Discover Publications. 79: National Solar Observatory (2). 80, 81: Art by Fred Holz. Butterfly diagram by Daniel C. Wilkinson, National Oceanic and Atmospheric Administration/National Geophysical Data Center. 83: USAF photo. 84, 85: Big Bear Solar Observatory, California Institute of Technology. 86, 87: NASA/Technology Application Center, University of New Mexico, Albuquerque. 89: The Naval Research Laboratory. 91-99: Art by Alfred Kamajian. 100, 101: George Cresswell, Commonwealth Scientific and Industrial Research Organization, Australia. 102: Initial cap, detail from pages 100, 101. 103: Jay Pasachoff, Williams College; High Altitude Observatory, Boulder, Colo. 104, 105: Hale Observatories, Palomar Mountain, Calif. (2). 106: University of Chicago, Office of News and Information. 108, 109: Soviet Embassy, courtesy *Sky & Telescope* magazine; NASA, Washington, D.C. (3); Ball Aerospace Systems Group. 111-113: Art by Matt McMullen. 114, 115: Art by Matt McMullen—graphs by Time-Life Books. 116: NASA, LBJ Space Center, Houston (SL4-143-4706)—NASA, George Marshall Space Flight Center, Alabama. 118: American Science and Engineering, courtesy Leon Golub, Smithsonian Astrophysical Observatory, Cambridge, Mass. 121: High Altitude Observatory, Boulder, Colo. (3). 124, 125: Art by Fred Devita, adapted from National Center for Atmospheric Research/National Science Foundation illustration. Inset art by Fred Devita. 126-131: Art by Fred Devita. 132, 133: Art by Fred Devita, adapted from "The Earth's Magnetotail," Edward Hones, Jr., © March 1986 by Scientific American, Inc., all rights reserved.

Time-Life Books Inc.
is a wholly owned subsidiary of
THE TIME INC. BOOK COMPANY

President and Chief Executive Officer:
Kelso F. Sutton
President, Time Inc. Books Direct:
Christopher T. Linen

TIME-LIFE BOOKS INC.
EDITOR: George Constable
Executive Editor: Ellen Phillips
Director of Design: Louis Klein
Director of Editorial Resources: Phyllis K. Wise
Director of Photography and Research:
John Conrad Weiser

PRESIDENT: John M. Fahey, Jr.
Senior Vice Presidents: Robert M. DeSena, Paul R.
Stewart, Curtis G. Viebranz, Joseph J. Ward
Vice Presidents: Stephen L. Bair, Bonita L.
Boezeman, Mary P. Donohoe, Stephen L.
Goldstein, Juanita T. James, Andrew P. Kaplan,
Trevor Lunn, Susan J. Maruyama,
Robert H. Smith
New Product Development: Yuri Okuda,
Donia Ann Steele
Supervisor of Quality Control: James King

PUBLISHER: Joseph J. Ward

Editorial Operations
Copy Chief: Diane Ullius
Production: Celia Beattie
Library: Louise D. Forstall

Computer Composition: Gordon E. Buck
(Manager), Deborah G. Tait, Monika D. Thayer,
Janet Barnes Syring, Lillian Daniels

Correspondents: Elisabeth Kraemer-Singh (Bonn),
Christina Lieberman (New York), Maria Vincenza
Aloisi (Paris), Ann Natanson (Rome). Valuable
assistance was also provided by Trini Bandres
(Madrid), Dick Berry (Tokyo), Elizabeth Brown
(New York), John Dunn (Melbourne), Christine
Hinze (London), Felix Rosenthal (Moscow),
Patricia Strathern (Paris), and Ann Wise (Rome).

VOYAGE THROUGH THE UNIVERSE

SERIES DIRECTOR: Roberta Conlan
Series Administrator: Judith W. Shanks

Editorial Staff for *The Sun*
Designers: Robert K. Herndon (principal),
Ellen Robling
Associate Editor: Kristin Baker Hanneman
(pictures)
Text Editors: Allan Fallow (principal),
Lee Hassig, Robert M. S. Somerville
Researchers: Mark Galan, Karin Kinney,
Mary H. McCarthy, Barbara C. Mallen
Assistant Designer: Barbara M. Sheppard
Copy Coordinators: Anne Farr,
Darcie Conner Johnston
Picture Coordinator: Ruth Moss
Editorial Assistant: Katie Mahaffey

Special Contributors: Sarah Brash, Gregory
Byrne, K. C. Cole, James Cornell, Ken Croswell,
James Dawson, Peter Kaufman, Laura Ost, Peter
Pocock, Chuck Smith, James Trefil, Karen and
Wallace Tucker (text); Sydney Baily, Vilasini
Balakrishnan, Craig Chapin, Edward Dixon, Dan
Kulpinski, Jocelyn Lindsay, Philip Murphy,
Eugenia Scharf, Elizabeth Thompson, Roberta
Yared (research); Barbara L. Klein (index).

CONSULTANTS
DEBORAH HABER, public-information officer at
the National Solar Observatory, Sacramento Peak,
Sunspot, New Mexico, researches the interaction of
solar magnetic fields with solar oscillations.

JOHN HARVEY is a member of the scientific staff of
the National Solar Observatory, Tucson, Arizona,
where he researches solar magnetic and velocity
fields.

FRANK HILL, a staff member at the National Solar
Observatory in Tucson who is actively involved in
the Global Oscillation Network Group *(GONG),* uses
solar-oscillation data to monitor the Sun's convection zone.

ROBERT F. HOWARD is an astronomer at the National Solar Observatory in Tucson and an editor of
the journal *Solar Physics.* He previously directed the
solar program at Mount Wilson Observatory in Pasadena, California.

JAMES C. LOPRESTO, a specialist in solar rotation,
helioseismology, and high-resolution solar spectroscopy, teaches at Edinboro University of Pennsylvania and is director of the university's observatory and Solar Observational Laboratory.

DONALD F. NEIDIG is an astrophysicist at the National Solar Observatory, Sacramento Peak, Sunspot, New Mexico. He specializes in solar flares and
solar activity forecasting.

JAY M. PASACHOFF directs the Hopkins Observatory at Williams College, Williamstown, Massachusetts, where he teaches. He is the author of text-
books on astronomy and other sciences.

C. T. RUSSELL participated as principal investigator on several NASA missions to explore Earth's
magnetosphere, the planets, and interplanetary
space. He teaches at the University of California,
Los Angeles.

RAYMOND N. SMARTT is deputy director for operations of the National Solar Observatory at Sacramento Peak, Sunspot, New Mexico. His areas of
expertise include instrumentation, optical design,
and coronal physics.

JACK B. ZIRKER, an astronomer and physicist at the
National Solar Observatory, Sacramento Peak, Sunspot, New Mexico, studies the Sun's corona, chromosphere, and photosphere. From 1976 to 1984, he
served as director of the observatory.

**Library of Congress Cataloging in
Publication Data**
The Sun/by the editors of
Time-Life Books.
p. cm. (Voyage through the universe).
Bibliography: p.
Includes index.
ISBN 0-8094-6887-5
ISBN 0-8094-6888-3 (lib. bdg.)
1. Sun—Popular works. I. Time-Life Books.
II. Series.
QB521.4.S86 1990
523.7—dc20   89-4468 CIP

For information on and a full description of
any of the Time-Life Books series, please call
1-800-621-7026 or write:
Reader Information
Time-Life Customer Service
P.O. Box C-32068
Richmond, Virginia 23261-2068